GIS Tutorial for Python Scripting

David W. Allen

Esri Press
REDLANDS|CALIFORNIA

Esri Press, 380 New York Street, Redlands, California 92373-8100
Copyright © 2014 Esri
All rights reserved.

Printed in the United States of America
19 18 17 16 15 14   1 2 3 4 5 6 7 8 9 10

Ask for Esri Press titles at your local bookstore or order by calling 800-447-9778, or shop online at esri.com/esripress. Outside the United States, contact your local Esri distributor or shop online at eurospanbookstore.com/esri.

Esri Press titles are distributed to the trade by the following:

In North America:

Ingram Publisher Services
Toll-free telephone: 800-648-3104
Toll-free fax: 800-838-1149
E-mail: customerservice@ingrampublisherservices.com

In the United Kingdom, Europe, Middle East and Africa, Asia, and Australia:
Eurospan Group
3 Henrietta Street
London WC2E 8LU
United Kingdom
Telephone: 44(0) 1767 604972
Fax: 44(0) 1767 601640
E-mail: eurospan@turpin-distribution.com

CONTENTS

PREFACE

Do you want to be an expert Python programmer? Well, so do I! That desire put me on a path of investigation and learning to find out everything I could about using Python with ArcGIS, from A to Z, and eventually led me to writing this book.

As a bit of a confession, I'm not the world's greatest Python programmer, but I overcome that by being the best GIS analyst I can be and knowing how to research the tools that I'm already familiar with.

My strategy is to write detailed pseudo code, research the tools I will need in ArcGIS for Desktop Help, test those tools, and then try out what I have learned. I often write code in small steps, debugging each part of the overall program before moving on to the next. As an example, I may write and debug a file collection process, add and debug a cursor to go through the files, and then add and debug the code to modify the files. I find this process much easier than trying to write massive amounts of code and then having to debug several processes and techniques that may be used all at once. This strategy makes me an excellent Python programmer in the realm of ArcGIS, which is what I want you to be!

Your work in the tutorials will reflect that process. After all, the end user will not see your code but will be amazed at the job it does. You can get your Python basics from the other books, and in this book, I will show you the tools and functions you'll want to use most in ArcGIS. Then you can work through a variety of hands-on tutorials and exercises that will have you writing pseudo code, creating scripts from scratch, and solving problems like a pro.

The other part of being a good programmer is learning to think like a computer: your steps should be clear and concise and follow a logical path. That clarity is sometimes contrary to how we function in the real world and to how we use language every day. As an example, let's say you are on a train in Germany, traveling to Wolfsburg, and a passenger makes an upward gesture and says to you in English, "Please, sir, to help me on the shelf on top of my head to put my bag." You would be able to interpret this sentence and understand that he wants help putting his bag in the overhead compartment. A computer, however, would not understand this sentence at all nor would it be influenced by any voice inflection or gestures. Its interpretation might be that you want to put the man on the shelf and then put the bag on top of his head. The sequence and understanding of the process is not right. The construction of detailed pseudo code will help you think through the process, ensuring that the instructions are clearly stated and that they follow a logical sequence. The more detailed the project, the more important the pseudo code. In fact, it is nearly impossible to write a Python add-in without pseudo code.

I regret that I cannot include every possible tool and technique that exists for Python programming—there are just too many. However, this book covers many of the more commonly used processes that are afforded to programmers in the ArcGIS world. When you master these processes, you will be able to look out into the Python world and begin to learn and understand the more advanced programming techniques that are available to enhance your code.

INTRODUCTION

A few Python books are now available that are specific to using Python in ArcGIS—with one problem. Most of them are not hands-on workbooks. These books describe the tools individually but do not go as far as showing you how to weave the tools together into scripts to accomplish a goal. It is the equivalent of showing you a picture of a hammer, describing how nails work (note which end is pointed), and then expecting you to build a house. The practical knowledge you need to become a programmer is included in this workbook.

With each of the tutorials, you should read the introduction, scenario, and description of the data. Following these sections is a section called "Scripting techniques," which outlines some techniques that you might want to use in your script. This section may also describe aspects of tools that you may not necessarily use but are useful to know. You should try writing your own pseudo code without looking ahead at the suggested solution, and then check and modify your completed pseudo code against it. If you have a substantially different process from what the book presents, work the tutorial as written so that you will learn the techniques included, and then feel free to go back and work the tutorial a second time using your own pseudo code. It will be great practice. Data for the book is available to download on the Esri Press "Book Resources" webpage, esripress.esri.com/bookresources. Click the appropriate book title, and then click the data link under "Resources" to download the exercise data. A 60-day trial of ArcGIS for Desktop software and extensions is available for readers at esri.com/trydesktop.

You may also find that some of the code is not as concise as hard-core programmers might write. This level of code is used, in part, to keep different tools and techniques separated and to make it easier for you to understand. Students sometimes have a problem looking at sample code that has several advanced techniques going on and trying to pick out and learn each technique separately. The simplicity also helps to make the code easier to debug. As you are learning, you may not always know where to start with the debugging process, and complex code will only make debugging more confusing. As you become more advanced, work on making your code more concise.

This book's tutorials are divided into five chapters. Chapter one goes over many of the components of Python and shows you how to use them in label expressions and field calculations in ArcMap. This chapter also introduces decision-making and condition statements and explains how they can all be used together. Chapter two has you start writing stand-alone Python scripts and script tools and using them in ArcMap and ArcCatalog. Chapter three works with the ArcPy mapping module used to control the map elements in your layouts. Chapters four and five cover the design and creation of more advanced features, such as Python toolboxes and Python add-ins.

As you go over these tutorials and exercises, take an extra moment to try the techniques on your own data, or at least on data that you are more familiar with. As with many subjects, the more you practice, the better you will become. It is also important to see how these techniques apply to your specific situation; that is what is going to ingrain these tools into your thought process and make you a good programmer.

Python code is edited with specialized programs called IDEs (integrated development environments), and there are several good IDEs for Python. When you install Python, it comes with an IDE called IDLE. In addition, you may want to install PythonWin, which ships with ArcGIS but is restricted as to the operating systems it can use. The examples in this book are created using an IDE called PyScripter, which is free to download and install and has many rich features that make using it appealing. You are welcome to use any IDE you like. The results will be basically the same. More information on these IDEs is available in appendix A.

Chapter 1

Using Python in labeling and field calculations

Introduction

As you begin working with Python as a programming language and start incorporating Python scripts into ArcGIS, you will find that there are many places where Python code can be used. This use may be as a small code snippet as demonstrated in this chapter or in fully developed programs as you will see in later chapters. For these first tutorials, take extra time to research the various Python and ArcGIS components and the structure of the code. As the projects become more complex, you will appreciate understanding the basics of this type of programming.

Tutorial 1-1 Python introduction and formatting labels

Python code can be used in places other than fully developed scripts. The Label Expression dialog box in ArcGIS allows you to insert code to control labels on your map.

Learning objectives

- Basics of Python
- Text formatting
- Variable manipulation

Preparation

Research the following topics in ArcGIS for Desktop Help:

- "What is Python?"
- "Building label expressions"

Introduction

Python is a powerful scripting and programming tool, but you need to know the basic rules of the game before you start playing. This tutorial presents a summary of the components most commonly used in ArcGIS. You can reference the Python documentation online at http://www.python.org and other Python reference books, such as *Python Scripting for ArcGIS* by Paul A. Zandbergen (Esri Press, 2013), for full descriptions and more advanced tools. Also, research the ArcGIS-related tool you will be using in ArcGIS for Desktop Help, where you will find descriptions of the tools and code samples that can be used to better understand the tool's usage.

Here are some basic rules for Python:

- Python code runs in a linear fashion—from top to bottom.
- Python includes variables, which can contain a variety of data types, including numbers, strings, lists, tuples, and objects (with properties).
- Variable types (e.g., numeric, string, list, date) do not need to be declared—Python determines the variable type based on the input.
- Variable names are case sensitive—"myFeatureClass" is not the same as "myfeatureclass."
- Either single or double quotation marks can be used when creating string-type variables—the Python code interpreter does not care, so "myFeatureClass" is the same as 'myFeatureClass.'
- Indentation in Python is important. Indenting is a way to group tools and operations into a set of code within your script, such as a code block associated with an if or while statement. Indentations are typically two spaces or four spaces; you can use tabs, but do not mix tabs and spaces.

The next few steps will let you practice some of these rules before you tackle the first tutorial.

1.   **Open your integrated development environment (IDE), and start a new script.**

The example shown in the graphic is a modified PyScripter template for ArcGIS that includes the name of the script, the author, a script description, the date of creation, and the license level that this script might require. Information on setting up this template in PyScripter is found in appendix A. Note that these lines are preceded by a hashtag, which denotes them as comments and not code that can be run.

```
#-------------------------------------------------------------------
# Name:        Rules Test
# Purpose:     Sample code to demonstrate Python
#
# Author:      David Allen
#
# Created:     02/17/2014
# Copyright:   (c) David Allen 2014
# License:     ArcGIS 10.2
#-------------------------------------------------------------------
```

2.   **Type the code as shown:**

```
myName = "David"
print "My name is " + myName + " and your name is " + yourName + "."
yourName = "Holly"
```

This code creates a couple of variables and prints them to the IDE code window. Note that the variable is created simply by using the equals sign, and the various parts are brought together in the print statements using the plus sign. This is called *concatenation*, which basically creates one line of text out of all the components.

If you were to try and run this code, you'd get an error. Why? Python runs these lines in order, from top to bottom. The print statement is run before the second variable is defined.

3.   **Change the order of the statements so that they will run correctly. Save the script for future reference, and then run the script.**

Remember that Python runs from top to bottom, so the lines of code must be in the correct order.

4.   **Type the code shown in the graphic to use different types of formatting to create four variables:**

```
streetName = 'Candy Lane'
addressNumber = 313
percentOccupied = 45.35
ownerName = "Brown, Charles"
```

Note the format of the variable names. The names are descriptive of what they contain, start with lowercase letters, and use uppercase letters to distinguish words within the name. This is called *camel case*. Although this format is not required, it is standard in the ArcGIS for Desktop Help sample code.

Two of the example variables shown are strings (text), and each of them uses a different style of quotation marks. Both styles can be used, and both are considered regular strings. Note that the numbers are also different. One has decimal points, and one does not, but both are still interpreted in Python as numeric. The IDE shows these lines in different colors so that you can tell the number variables from the string variables.

Expressions can be used to concatenate the strings and to perform math on the numbers. The graphic in step 2 shows an example of concatenating string variables with the plus sign. Numbers can be concatenated into these types of sentences as well, but numbers must first be converted to strings using the string formatting method, .str(). Because the IDE colored the numbers differently, you can easily tell when a conversion is necessary.

5.   **Type the code shown in the graphic, and then run your script to see the results.**

```
• print "The owner of " + str(addressNumber) + " " + streetName + " is " + ownerName + "."
• print "The unoccupied area is " + str(100 - percentOccupied) + "."
```

The strings are concatenated together, and the number is added once it is converted to a string. Note that in the second line, there is some math occurring inside the string conversion function. This calculation is fine as long as the result is converted to a string. Also, pay attention to where the extra spaces are added to the text to make the sentence appear correct when printed.

```
The owner of 313 Candy Lane is Brown, Charles.
The unoccupied area is 54.65.
>>>
```

It is also possible to slice characters from a string variable. Each character in the variable is automatically assigned an index number, starting at the left with zero (0). You can count over to the characters you want, and then slice those characters from the string. To slice characters, add square brackets at the end of the variable ([]), and then inside the brackets, add the starting index number, a colon, and the ending index number.

6.   **Type the print statement shown in the graphic to slice only the street name from the streetName variable.**

```
• print "The street name is " + streetName[0:5] + " and is designated as a " + streetName[6:] + "."
```

The first slice gets all the characters starting at index number 0 over to but not including index number 5. The second slice gets all the characters starting at index number 6 and over to the end. The indexes can also be counted with negative numbers from the end of the string. The word *Lane* could also have been sliced using this statement, which gets all the characters starting four back from the end and proceeding to the end, as shown:

```
• print "The street name is " + streetName[0:5] + " and is designated as a " + streetName[-4:] + "."
```

A common use of the negative slice is to remove the file extension from the end of a file name. The example in the previous graphic, using -4, would remove .shp from the end of a file name, regardless of its length.

Another type of variable is a list. List variables can contain many values and are used extensively in ArcGIS to hold lists of feature classes, file names, and workspaces. The individual values within the string are accessed by using an index number. Each value is given an index number, starting at 0. For example, a variable with eight list items would have index numbers from 0 to 7.

7. **Type the code shown in the graphic to assign a list to a variable, and then print one of the values using its index number.**

```
listNames = ["David", "Candy", "Holly", "Timmy"]
print listNames[2]
print listNames[3]
```

Remember that the first value is given index number 0, so this code will print the names Holly and Timmy.

These are simple examples of creating and manipulating variables. More complex manipulations follow in other sections of the book. Try some of these things in ArcMap.

8. **Close the IDE you've been using and save the file for future reference.**

These text manipulations can be used in various parts of ArcGIS, and this tutorial examines using them in a labeling expression. The interface you use for labeling features in a map layout will recognize Python script and allow you to do formatting and make on-the-fly changes to the text that you may be using from a layer attribute. This feature may save you a lot of time when the attribute values are not formatted exactly as you like or when your data has to meet a standard that is different from other datasets you may be using.

Scenario

A standard ArcMap layout is used by your department to review the owners of properties in the fictitious City of Oleander, Texas. The formatting is not the best, but it is functional for your internal use. Recently, the city manager asked you to make more of this type of data available through online maps, and the formatting is not appropriate. You should explore some ways to use Python scripting to make the text more presentable.

Data

The data is the parcel and ownership data for the City of Oleander, Texas, a small community in the Oleander/Fort Worth Metroplex (O/FW). A map document is set up to display data for each property from the owner name field.

SCRIPTING TECHNIQUES

The Label Expression dialog box has a Python function built in to make it easier to use the code. A function is basically a set of code that can receive one or more values and return values back to the program that called it. In the case of the label expression, the built-in function is named FindLabel() (which you should not change). This function appears on a line starting with *def*, meaning that it defines the function. The function may also have one or more field names at the end appearing in parentheses. These field names are passed to the code of the function so that you can use and manipulate their values. Finally, a return statement returns a value to the program that called the function, which in this case is the label expression. For labels, make sure to return a single value.

The Label Expression dialog box lets you use a variety of Python text formatting methods, including the following:

- .capitalize()—makes the first character of the string a capital letter
- .find(X)—finds the specified character in the string
- .isdigit()—returns the value of True if the variable is alphanumeric
- .lower()—makes the entire string lowercase
- .lstrip()—strips any spaces from the left end of the string
- .replace(X,Y)—replaces one character with another in the string
- .rstrip()—strips any spaces from the right end of the string
- .strip()—strips spaces from the start and end of a string
- .title()—capitalizes the first character of every word in a string
- .upper()—makes the entire string uppercase

Check your Python reference books for more string and numeric handlers for variables. Some are obscure, but you never know when they will be useful.

Use Python in the Label Expression dialog box

1. Using the information provided, write the sequence of the calculation in nontechnical language, as shown in the following list. Then use this sequence to decide the formal structure of the code. The goal is to use Python string-handling techniques to dress up the labels on this map.

- Change the text to read as a regular name (first, then last).
- Add an ampersand (&) if a couple owns the house.
- Change the text to upper- and lowercase rather than all uppercase.

This informal description of the processes your code will complete is the pseudo code. Even simple pseudo code is useful, and as you tackle more complex projects, your pseudo code will reflect this complexity.

2.   **Start ArcMap, and open the map document Tutorial 1-1 from the location where you installed the book's sample data and exercises.**

LAWSON, JIMMY ETUX MARY	NORRIS, JACK	RAWLINGS, MAURY KIM	SINGLES, BRANDON S ETUX LESLIE	LARSEN, LINDA & LANCE
CRAMER, ROY ETUX NOVA	RYAN, STEVEN J ETUX ELIZABETH	GRUBBS, RYAN L	OTWELL, BOBBY COY	NGUYEN, NAM D ETUX TAMI

The map in the graphic shows the Elm Fork subdivision with a label on each property that displays the owner's name. Although this format is acceptable, the map would look better if the text were formatted to show the first name followed by the last name. The current format of the data is "last name–comma–space–first name(s)." This data can be sliced into separate pieces to show it as "first name(s)–space–last name." This slice involves using the .find() method on the string, which returns the index number of the character you will search for, and then using that location in a slice operation. If you can find the location of the comma, everything before the comma is the last name, and everything two characters past the comma (remember the space) is the first name(s).

3.   **Open the properties of the Elm Fork Addition layer, and click the Labels tab. Then click the Expression button. It shows the expression as [OwnerName]. At the bottom of the Expression box, click the Parser arrow, select Python, and then select the Advanced check box in the middle of the dialog box on the right side.**

This step automatically enters the first few lines of Python code to define the FindLabel() function. At the end of this line is a colon (:). All the code that is to be evaluated as part of this function will appear below this line and be indented, ending with a return statement that sends a value back to the FindLabel() function and ultimately onto the map. The standard indentation in Python is four spaces, but the code here uses only two. Next, create a new line of code, indent the code two spaces, and use the .find() method to determine the index number of the comma in each value.

4. **Place your cursor after the colon, and press Enter. Add two spaces, and then type the following:**

rawName = [OwnerName]

It is not always necessary to put the field name into a variable, but it will make the process easier to understand.

5. **Press Enter, add two spaces, and type the following:**

commaNum = rawName.find(",")

The value of commaNum is now equal to the index number of the comma. See if you can write the code to format the string correctly, and store it in a variable named **formatName**. Here is what needs to happen:

Slice out everything starting at two characters past the comma over to the end of the string, add a space, and add the slice of characters from the start of the string over to but not including the comma.

6. **After the line of code, press Enter, and add two spaces. Type the code to perform the field formatting. Then replace the expression [OwnerName] in the last line with** formatName**, as shown. Click OK, and close the layer properties to see the result.**

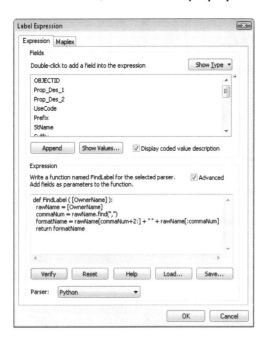

Notice how you can do math in the slice command. A few more things still need to be corrected. The ownership files use the Latin abbreviation ETUX (et ux) in front of a spouse's name. Replace that with an ampersand using the .replace() method.

7.   **Open the layer properties, and modify the expression by adding the .replace() method at the end of the rawName statement, as shown:**

```
rawName = [OwnerName].replace("ETUX","&")
```

8.   **Add the .title() method at the end of the return statement to make the names appear in upper- and lowercase, as shown:**

```
return formatName.title()
```

The resulting labels show the first and last names in upper- and lowercase, along with ampersands in place of the combined wording ETUX.

9.   **When your code matches the graphic, click the Save button, and save the code you have written to a calculation (.cal) file for later reference. Then click OK to run the script and perform the calculation.**

Jimmy & Mary Lawson	Jack Norris	Maury Kim Rawlings	Brandon S & Leslie Singles	Linda & Lance Larsen
Roy & Nova Cramer	Steven J & Elizabeth Ryan	Ryan L Grubbs	Bobby Coy Otwell	Nam D & Tami Nguyen

10.   **Close the map document.**

Exercise 1-1

Open the map document Exercise 1-1, a map of the subdivisions in Oleander with their names displayed. The city manager needs to show this map to another city and wants you to dress it up a little. First, the names should not have .PDF at the end, and second, the names should be in upper- and lowercase with the first letter of each word capitalized.

Write the steps needed to accomplish this task, including the Python code to use for the labeling. Then apply the code to the labeling expression in the map to reformat the text.

Tutorial 1-1 review

This code performed a lot of string handling with different variables. Note how you are able to find specific characters in a variable, store the index number, and use that number to slice the string in different ways. There may be times when this gets tricky, such as when an address number has an apartment or unit number that is a letter character (e.g., 304 A Pine St.). Looking for the first space would not give you the correct address number. You could look for the instance where a character stands alone—in other words, any single character with a space on both sides. To give a more complicated example, what if you had single-letter street names, such as D Street or P Street? The stand-alone character is the street name and not part of the address number. To make it even more confusing, Galveston, Texas, has half streets, such as P½ Street. This situation may require some complex Python coding to solve.

Study questions

1. Will finding the first space always identify the address number? Could you find all the spaces? Write some code to find the first stand-alone character in a string (a single character with a space on both sides).

2. Can strings and numbers be combined in Python? Write code to concatenate a text string, such as "Miles to go are," with a numeric variable, such as 102.

3. Give examples of when you might use the .upper(), .title(), and .lower() methods.

Tutorial 1-2 Decision making in the Label Expression dialog box

Labels can be simple displays of data from an attribute table, or with some Python coding, labels can be used to show values that are not in the table.

Learning objectives

- Defining functions
- Using if-elif-else logic for decision making
- Changing the label display text

Preparation

Research the following topic in ArcGIS for Desktop Help:

- "Using If-then-else logic for branching"

Introduction

In tutorial 1-1, you learned how to use Python code to control labels in a map document using various text-handling techniques. Those techniques involved simply arranging and reformatting the existing values of the fields used for labeling. You can also create labels that use entirely different text strings from the fields—it all depends on what you send back to the label expression with the return command.

In this tutorial, you will apply decision-making techniques to the labels in ArcMap. You will be able to pull the value out of the field and use it to determine what the label should contain.

Scenario

A file has been provided by the planning office of Oleander. This file needs a map to display at City Council meetings that shows the general zoning categories of Oleander. You have a good dataset for this, but the zoning is shown as code rather than a text description. You must use some decision-making tasks to turn the codes into the appropriate labels.

Data

You are provided with the zoning map for the City of Oleander containing a layer with the zoning district polygons named General Zoning Districts. This layer has a field named Code, which contains a coded value for zoning. The codes are as follows:

RES = Residential
MF = Multi-Family
SPEC = Special District
C = Commercial
I = Industrial

SCRIPTING TECHNIQUES

You will need to build a condition statement using the if statement. This book covers if statements in more detail in later chapters, but for now, here is a quick introduction.

Every programming language has some version of the if statement to perform branching based on a condition being true, and Python is no exception, providing one of the easiest if formats. The first if statement tests a condition that can evaluate to either True or False, along with a set of code to run if the statement is true. This statement can be followed by any number of elif statements to run if the condition is tested to be false—and the elif condition statements are all evaluated in sequence until one is true. Finally, an else statement is provided with no condition to be met and is run if all the other conditions are false. A basic decision-making statement looks like this:

```
#Check to see which statement equals "RUN ME"
if statement1 == "RUN ME":
    # run this code
    return Value
elif statement2 == "RUN ME":
    # run this code
    return Value
elif statement3 == "RUN ME":
    # run this code
    return Value
else:
    print "Nothing is ready to run"
    return Value
```

Note that the evaluator is a double equals sign (check your Python reference for other evaluators such as the greater than symbol or the lesser than symbol), and the code for each if, elif, or else statement is kept indented until the next statement. This format designates which lines of code to run if the statement evaluates to True. The return statement tells the code what to send back to ArcMap—in this case, the text for the label.

Use if-elif-else statements in the Label Expression dialog box

1. Write the pseudo code for this project:

- Get the code value from the field.
- Determine what text to use for each code value.
- Return the text string to the label expression.

2. Start ArcMap, and open the map document Tutorial 1-2. Right-click the General Zoning Districts layer, and open the properties. If necessary, click the Labels tab as shown, and set the Label field to CODE.

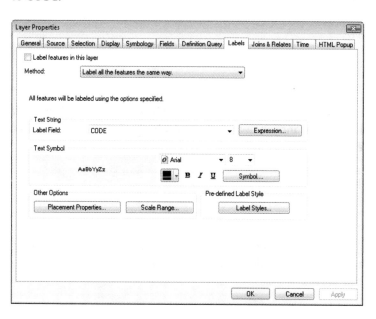

3. Click the Expression button. Do you remember how to set this to accept Python code? (Hint: set the Parser to Python and click Advanced.)

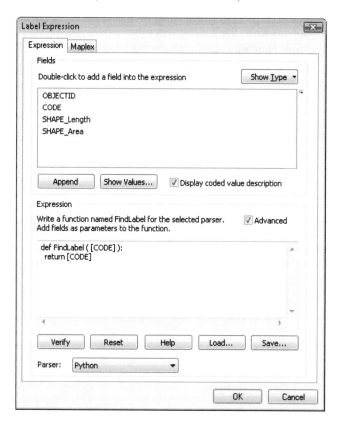

4. The framework is ready for the code to be entered. First set a variable to equal the zoning code that is brought into the script by the function. Add two spaces before the variable name to keep the indentation, as shown:

```
def FindLabel ( [CODE] ):
  zoneCode = [CODE]
  return [CODE]
```

5. Construct the if statement. Remove the line "return [CODE]" because this will be replaced within the if statement. Add a new line, indent two spaces, and add the if statement, remembering to add the colon at the end of the line, as shown:

```
def FindLabel ( [CODE] ):
  zoneCode = [CODE]
  if zoneCode == "C":
```

6. Add another line, and indent four spaces. Everything that keeps the four-space indentation will be evaluated when this if statement evaluates to True. Add a return statement to tell ArcMap what to use as a label, as shown:

```
def FindLabel ( [CODE] ):
  zoneCode = [CODE]
  if zoneCode == "C":
    return "Commercial"
```

Next, add code to handle what happens when the if statement is not true—you will set up the next label. Python makes checking multiple conditions easy with the use of an elif statement. Remember that this statement also gets a colon at the end of the line, and everything after the colon that is kept to a four-space indentation will run if the condition is true.

7. Add another line, indent two spaces, and type the elif statement; then add another line with a four-space indentation and the return statement.

```
def FindLabel ( [CODE] ):
  zoneCode = [CODE]
  if zoneCode == "C":
    return "Commercial"
  elif zoneCode == "I":
    return "Industrial"
```

Your turn

*Add the rest of the code to test each of the remaining conditions. Remember that the last line can be an else statement with no condition. Try it from your own notes first before looking at the code in the next image. (**Hint:** make sure to watch your case on variables; use the == evaluator; put a colon at the end of each line with an if, elif, or else statement but not after the return statement; and watch your indentations carefully).*

```
def FindLabel ( [CODE] ):
  zoneCode = [CODE]
  if zoneCode == "C":
    return "Commercial"
  elif zoneCode == "I":
    return "Industrial"
  elif zoneCode == "MF":
    return "Multi-Family"
  elif zoneCode == "RES":
    return "Residential"
  else:
    return "Special District"
```

8. **Click OK and then OK to run the code and create the labels.**

If you have an issue, go back over the code, and pay particular attention to the things noted in the "Your turn" hint.

You will see that your code has turned the zoning codes into a more descriptive label. It is important to remember that these labels are generated on the fly and used for this map only; these labels are not stored anywhere in the attribute table.

Exercise 1-2

The other map turned out so well that the planner wants a similar map for the comprehensive land development data. Open the map document Exercise 1-2, and use Python code in the label expression to turn the following USE_CODE values into descriptive text:

A1 = Single Family
B1 = Multi-Family
F1 = Commercial
F2 = Industrial
PRK = Parks
ROW = Right-of-Way
TX10 = Special Texas District

Write your code before attempting to type the expression in the Label Expression dialog box.

Tutorial 1-2 review

The if statement in Python is one of the easiest to use in the programming world. For any set of conditions, continue adding elif statements until you have handled all the possible conditions. It is good practice to put an else statement after the last elif statement just to handle a situation that is not the norm or that you might have forgotten about or to handle the last known condition without having to use a condition statement.

If statements can also be nested, but this is only done to test two entirely different sets of conditions. For instance, your first if statement might be to identify the land-use code, and you nest an if statement within that statement to determine whether the particular parcel fits within a certain range of acreage. However, you would not need to use nesting to test for more land-use conditions as you would with other programming languages. These conditions could easily be handled with additional elif statements, one for each additional land use.

Another interesting note with if statements is that the condition need only return a value of True or False. If you are testing a variable that equates to True or False, you need only put the variable as the condition for the if statement. For instance, the .isDigit() function returns the values of True for a variable containing alphanumeric characters and False if the variable is empty. If you retrieved a variable named Name that contained characters, the code myVariable = Name.isdigit() would set myVariable to either True or False, depending on whether the field had an alphanumeric value. This situation would make the if statement look like this:

```
myVariable = Name.isdigit()

if myVariable:
    print "This field has characters!"
else:
    print "This field is empty."
```

The if condition statements can also test for more than one variable at a time, just like a query statement in ArcMap. The only trick is that there must be a combination of values that equate to True, or the code will never run. For instance, if you needed to find all the 40-inch or larger PVC pipe, the code would look like this:

```
if PMaterial == "PVC" and PSize > 40:
    print "Found a big plastic pipe!"
```

It does not matter that one condition tested a string-type value and the other tested a numeric value. Be careful with the condition though, because an incorrect AND or OR can send your code in an unwanted direction. If you are unfamiliar with the AND and OR handlers, check them out in ArcGIS for Desktop Help, and try practicing the statement as a definition query in ArcMap.

Study questions

1. When would you nest if statements, and when would you rely on elif statements? Write the code to find parcels that have a land-use code of A1, and note whether the acreage is less than two acres, from two to five acres, or more than five acres. Add additional code to find parcels with a land-use code of B1, and perform the same test for area.

2. How would you format a complex condition statement? Write the code to find employees over 50 who are retired if their age is stored in a field named *currAge*, and the field noting retirement status is a true/false field named *statusRetired*.

3. Where can you find more Python methods that deal with string and numeric variables? Write code to find all the employees with a last name containing more than 10 characters if their last names are stored in a field named *LastName* (or else it will not fit on the new engraved name tags).

Tutorial 1-3 Using Python in the Field Calculator

Snippets of Python code can be used in the Field Calculator to perform complex functions, including any of the text formatting commands shown in the Label Expression dialog box.

Learning objectives

- Text formatting with Python
- Using Python code blocks
- Concatenating text values

Preparation

Research the following topics in ArcGIS for Desktop Help:

- "Calculate field examples"
- "Fundamentals of field calculations"

Introduction

The first two tutorials show how to use Python code to alter the labeling in an ArcMap map document. This labeling made the maps look great, but remember that the changes occurred only in the labels and were not stored anywhere permanently.

To make a more permanent change to your data, you can use Python to calculate the values in fields. You would normally use the Field Calculator and a simple expression to set a field value, but there are limits to what you can accomplish with the simple expressions that are allowed. By using Python code in the Field Calculator, you can store the results for future use by yourself and others.

Scenario

You received some data from the Fire Department, and it would like you to format the addresses so that its analyst can geocode them and make a simple presentation map. To do the geocoding, the address needs to be in a single field, and right now that information is parsed out into five fields. You must write an expression in the Field Calculator to create a single address with the components in the right sequence, and without extra spaces.

Data

You are provided a file named FireRuns2010 with the calls for service data from the Oleander Fire Department. The file contains a variety of data about the type of call, the time it was received and dispatched, and which unit responded. The file also contains address information that has been split into five fields:

addNum = address number
stPrefix = street prefix
stName = street name
stType = street type
suffDir = street suffix direction

SCRIPTING TECHNIQUES

Using Python code in the Field Calculator is a little trickier than the labels expression because ArcMap does not automatically create the code for the function. You need to set the parser language to Python, and then select the Show Codeblock check box. This will expose two empty boxes where you will type your code.

The lower box, called the Expression box, will look familiar. This is the box that you normally work with for simple calculations and that you used to hold your label formatting code. Instead of using a regular statement, you will have the expression call a function from the code box described below. The syntax is to name the expression, which can be anything you like, but common practice is to begin the name with a lowercase *fn* to denote a function. Then in parentheses, add all the fields from the attribute table that you will be using in your code. It does not matter how many you use, and they will be separated by commas. You can add them by double-clicking the field name in the Fields list.

All the code work is done in the Pre-Logic Script Code box, or code box. The first line defines the function you named in the Expression box. It then has a set of parentheses and contains a variable name for each of the fields that you are sending to it. For instance, if your statement in the Expression box has five fields coming in, the function in the Pre-Logic Script Code box must set up five variables to accept them. The function ends with a colon, and all the code that follows must be indented. There is no automatic indenting here, so you must manually add spaces for indenting and then keep track of them.

Next, type all your Python code to process the data and perform the calculation, with the last line being the return statement. This code sends a value back to the Expression box, which is saved to the field. As with the label expressions, the use of if statements may result in several return statements.

Debugging this code can be problematic, but it is best to start by checking the indent levels. Then go back over the syntax of the code.

Use Python in the Field Calculator

1. Write pseudo code to describe the process for reformatting the field value and storing the result:
  - Concatenate the values for address number, street prefix, street name, street type, and street suffix direction into a single value.
  - Store this value in the empty field full_address.

2.   Start ArcMap, and open the map document Tutorial 1-3. If necessary, switch to the List By Source view in the table of contents. Open the FireRuns2010 table.

3.   Right-click the field full_address, and click Field Calculator. A warning about editing outside an edit session appears. Click Yes to continue.

It can be dangerous to edit outside an edit session in ArcMap because there is no undo option. If you make a mistake, it is permanent. In this case, you are populating an empty field, so there is no risk of destroying any critical data by calculating the field.

4.   In the Parser box, click Python. Below the Fields list, select the Show Codeblock check box.

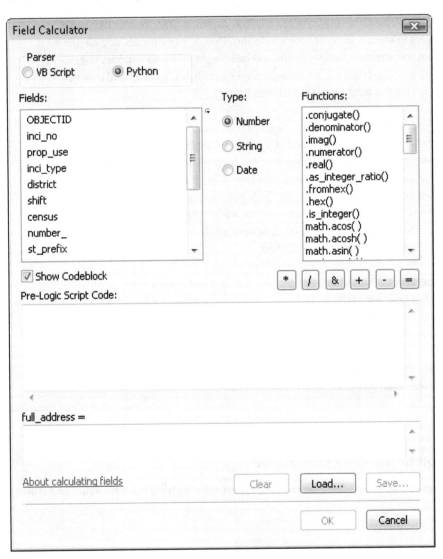

In the Expression box, you will create a name for the function, such as fnAddress, and write the Python code to perform your operation. The name here can be anything you like, but standard convention is to start functions with a lowercase *fn* to identify their type and add a descriptive name in uppercase.

5. In the Expression box under "full_address =," type the line shown in the graphic. Type the function name, and double-click the field names to enter them. Be sure to add commas between the field names.

full_address =

```
fnAddress( !number_!, !st_prefix!, !street!, !st_type!, !st_suffix_dir!)
```

Double-clicking the field names to add them to the function will automatically place the proper characters around the field name. These may be quotation marks, exclamation points, or square brackets, among others, depending on the data source. It is hard to know exactly which characters will be needed, so using the double-click method ensures that the correct characters are used.

6. In the Pre-Logic Script Code box, type the following code:

Pre-Logic Script Code:

```
def fnAddress(addNum,stPrefix,stName,stType,suffDir):
```

This defines a function named fnAddress and accepts each of the fields listed in the Expression box. **Note:** This is the format for all code that you will ever write in the Field Calculator. The Label Expression dialog box names a function, followed by all the fields that will be used in the script. Then the Pre-Logic Script Code defines the function beginning with *def* and accepts each of the fields into a Python variable that you name.

7. Write the Python code to format and concatenate the fields into a single string. Remember to account for a space between the values. When you have the code entered, click OK to perform the calculation. If you have difficulties, check your code against the graphic, but try writing the code on your own first.

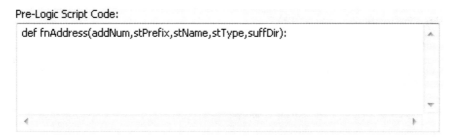

Pre-Logic Script Code:

```
def fnAddress(addNum,stPrefix,stName,stType,suffDir):
  formatAddress = str(addNum) + " " + stPrefix.strip() + " " + stName.strip() + " " + stType.strip() + " " + suffDir.strip()
  return formatAddress.title()
```

This worked pretty well, but there is still a problem. If the stPrefix string is empty, an additional space is added to the output string. Also, extra spaces are added at the end of the output string if suffDir is empty. Using what you know about if statements, try writing the code that will test for this field being empty, and then handle the case of what to do if it is empty. It should have this logic (add this to your pseudo code):

- If the field stPrefix is empty, write the concatenation without this field included.
- Add a command at the end of the output to strip off blank spaces.
- If the field stPrefix is not empty, write the same concatenation as before with the command added to strip blank spaces from the end of the string.

You should also investigate other Python formatting tools to remove any blank spaces from the values and perhaps to control the capitalization.

8. **Right-click the full_address field, and open the Field Calculator dialog box. Type the code you wrote—making sure to use the correct indentations with the if statements. (Hint: indent two spaces for every command after the def statement and four spaces for every command to run with the if statement.) Click OK to test it—the completed code looks like this:**

Pre-Logic Script Code:

```
def fnAddress(addNum,stPrefix,stName,stType,suffDir):
  if stPrefix == "":
    formatAddress = str(addNum) + " " + stName.strip() + " " + stType.strip() + " " + suffDir.strip()
  else:
    formatAddress = str(addNum) + " " + stPrefix.strip() + " " + stName.strip() + " " + stType.strip() + " " + suffDir.strip()
  return formatAddress.title()
```

Note the use of indentations to distinguish the commands for the different parts of the if statement. This example also includes the .strip() and .title() functions to help with the text formatting.

You can see that complex scripts can be developed for use in the Field Calculator. The format for any script you write will be the same: name a function in the Expression box along with the fields you will be using in the script, define a function in the Pre-Logic Script Code box with a Python variable for each field named in the function, and write the code to do your processing.

Exercise 1-3

The chief would like to export this data into another program for analysis, but there must be a field describing which station responded. The field district has a number that designates the station code, but the chief would like it in the format "Station 1" instead of the code.

Add a text field to the table FireRuns2010, and name it **Station.** Then write a script in the Field Calculator that will populate the field with the appropriate text:

151 = "Oleander Station 1"
551 = "Oleander Station 1"
152 = "Oleander Station 2"
552 = "Oleander Station 2"
153 = "Oleander Station 3"
553 = "Oleander Station 3"

Anything else should be made equal to "Outside Station."

As a bonus, add the shift code at the end of each value. For instance, for 551 shift B, the output would read "Oleander Station 1 – Shift B."

Tutorial 1-3 review

As you can see, the code in the Field Calculator can get complex. The trick is to maintain the indentations because the code block box does not handle indent levels automatically, as a good IDE would. It is sometimes good practice to build these statements in your IDE using some preset dummy data variables to test the syntax and set the indentations correctly. Then you can copy and paste the code to the Field Calculator.

The same rules that apply to if statements in the Label Expression dialog box also apply to if statements in the Field Calculator dialog box. Follow these rules carefully, and test any condition statements to make sure that they will not send your code out of control and that they have an instance that equates to True.

In both the Field Calculator and Label Expression dialog boxes, you are required to manage your own indent levels. Because of this, it is good practice to test your code in an IDE first for syntax and indentations before placing it in the Field Calculator. It is also advisable to calculate values into new, empty fields rather than to calculate values over existing values. If something is wrong in your code, you will destroy the original data values. These values would be impossible to recover if you were calculating outside an edit session.

This tutorial includes condition statements using string values. If the fields or values being compared do not match in case (uppercase, lowercase, or a combination thereof), the values may not equate to True, even when they are the same except for case. In this instance, you can use one of the string formatting functions to force the case to match before performing the testing.

Study questions

1. Could these same scripts be used to display labels on a temporary basis rather than storing the output string in a field? Write a code example to complete the exercise as a label statement (if you think it can be done).

2. What other string formatting statements are available for use in Field Calculator scripts? Write code to compare name fields from two different tables if the first is *Name* and stores a value in upper- and lowercase letters and the second is *EmpName* and stores a value in all uppercase letters.

3. Besides using double equals signs (==) for "is equal to," what other evaluators are available in Python? Write code to find pipe sizes starting at 8 inches and going up to 12 inches.

Tutorial 1-4 Decision making in the Field Calculator

Complex operations in the Field Calculator can include advanced Python math functions and if-elif-else logic.

Learning objectives

- Writing a Python code block
- Using if-elif-else logic
- Text formatting

Preparation

Research the following topic in ArcGIS for Desktop Help:

- "Using if-then-else logic for branching"

Introduction

In tutorial 1-3, you learned how to perform various text formatting techniques in the Field Calculator. The Field Calculator allows you to perform a variety of math functions as well. Python has many math expressions built in, such as adding, subtracting, multiplying, dividing, exponentials, and square roots, but Python also has a math module that can be imported to add higher-level math operators. This scenario uses simple math operators, but you can explore other Python code references for more options.

Scenario

The city engineer is preparing to do a flow rate study on the sewer system data. It is a gravity flow system, and the flow rate in gallons per minute (gpm) of wastewater must be calculated for each pipe size. In addition, she wants you to include a drag coefficient for the different pipe materials because some types of pipe are not flowing at the optimum rate due to friction or buildup in the pipes. Because you do not know the slope of the pipes, calculate the flow rate assuming a 1 percent grade, and she can calculate a more accurate flow rate when better slope data is acquired.

Your script must find the pipe size, get the flow rate for that size, and multiply the flow rate by the drag coefficient of the pipe material. The flow rates are as follows:

4 in. = 30 gpm
6 in. = 70 gpm
8 in. = 175 gpm
10 in. = 280 gpm
12 in. = 410 gpm

The drag coefficients for the different materials are as follows:

Ductile iron = 0.82
Reinforced concrete = 0.88
Vitrified clay = 0.92
High-density polyethylene = 0.97
Polyvinyl chloride = 0.97

Notice that even the best pipe, with a coefficient of 0.97, does not allow for wastewater to flow at the maximum rate. Because the pipe must maintain an air gap to allow the wastewater to flow freely, the maximum theoretical flow rate is never achieved.

Data

The data is the sewer utility data for the City of Oleander. The pipeline database already has an empty field in which you will calculate the flow rate. In addition, there is a field containing the pipe size and another one containing a description of the pipe material. A definition query has been applied to limit the pipe sizes to less than 14 inches to save time typing a lot of code. In reality, this process would be done on the entire dataset.

SCRIPTING TECHNIQUES

The first few tutorials used various Python controls to manipulate strings, but there are just as many available for mathematical functions. The common controls are addition (+), subtraction (-), multiplication (*), and division (/), but a variety of more complex functions exist, such as exponentials and square roots. A double-asterisk (**) operator is used for an exponential, so 2 to the power of 4 would be 2**4. Seven squared would be 7**2 (seven to the second power).

Python also has a separate math module that can be referenced from your scripts to do more scientific calculations, but these calculations are not addressed here.

Make decisions in the Field Calculator

1. Try writing your pseudo code for this project on your own before referring to the description shown:

```
# Get the pipe size and pipe material from the database
# Use the pipe size to determine the correct flow rate
# Find out the pipe material
# Perform the calculation
```

This pseudo code shows a general outline of the process.

2. Think about each of these steps, and determine what type of Python scripting you may have to write to accomplish the goal, and add that description to the pseudo code. You can be more specific in this step because the next step requires writing the actual code.

```
# Get the pipe size and pipe material from the database.
#    1. Define a function to bring the fields PSIZE and MATERIAL
#       into the Field Calculator.
#       Create two variables to hold these values.
# Use the pipe size to determine the correct flow rate.
#    2. Use an if statement to determine the pipe size.
#       There will be an if, then an elif for each pipe size in the list.
# Find the pipe material.
#    3. Nest an if statement in the pipe size if statement to set the
#       drag coefficient.
#       There will be an if, then an elif for each material type.
# Perform the calculation.
#    4. Perform the calculation, and store the result in a variable.
#       Use a return statement to send the results back to the function
#       and set the field value.
```

3. Start ArcMap, and open the map document Tutorial 1-4. This is the sewer utility data for Oleander, zoomed in on a small area. Open the attribute table for the Sewer Lines layer, and note the fields in the table that you will be using in this process.

PSIZE	MATERIAL	Year of Construction	Comments	Shape_Length	FlowRateGPM
6"	Ductile Iron	1992	Oleander	8.527382	<Null>
8"	Ductile Iron	1969	Oleander	339.713942	<Null>
6"	Ductile Iron	1985	Oleander	166.341481	<Null>
6"	Ductile Iron	1985	Oleander	71.561806	<Null>
6"	Ductile Iron	1998	Oleander	74.754366	<Null>
10"	Ductile Iron	2006	Bedford	238.383305	<Null>
12"	Ductile Iron	2006	Oleander	147.638852	<Null>
8"	Ductile Iron	2006	Oleander	130.098232	<Null>
10"	Ductile Iron	2008	Oleander	216.881243	<Null>
10"	Ductile Iron	2009	Oleander	36.19677	<Null>
12"	High Density Polyethelene	2009	Oleander	409.401769	<Null>
12"	High Density Polyethelene	2009	Oleander	304.087048	<Null>
8"	High Density Polyethelene	2009	Oleander	257.647263	<Null>
12"	High Density Polyethelene	2002	Oleander	20.836189	<Null>
12"	High Density Polyethelene	2002	Oleander	99.713169	<Null>
21"	High Density Polyethelene	2002	Oleander	38.267902	<Null>
8"	High Density Polyethelene	2010	Oleander	207.353537	<Null>
8"	High Density Polyethelene	2010	Oleander	378.480221	<Null>
8"	High Density Polyethelene	2010	Oleander	272.062532	<Null>
8"	High Density Polyethelene	2010	Oleander	128.980094	<Null>
8"	High Density Polyethelene	2010	Oleander	501.22632	<Null>
8"	High Density Polyethelene	2010	Oleander	160.566632	<Null>
8"	High Density Polyethelene	2010	Oleander	173.612591	<Null>

(0 out of 4218 Selected)

Sewer Lines

4.   Right-click the FlowRateGPM field, and open the Field Calculator dialog box. Click the settings to allow for the entry of Python code, and define a function to accept Python code as described in step 1 of your pseudo code. Refer to tutorial 1-3 for a reminder of how to do this. The function should pass the PSIZE and MATERIAL fields to the code block (in each step, try out your own code before referring to the graphics).

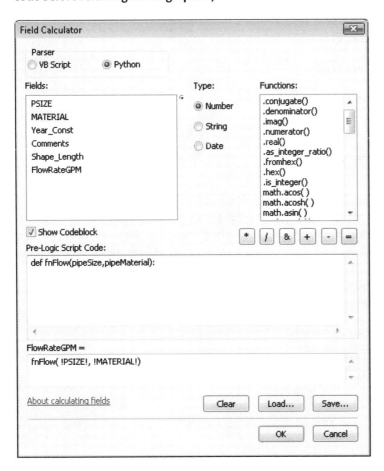

5.   Construct the if statement as outlined in step 2 of your pseudo code. Lay out the entire structure to determine pipe size before thinking about the steps to find the pipe material. Remember to control your indentations. The return statements return a blank value and are here just as placeholders. Note: The displays of code shown in the graphic are from an IDE for legibility. You may want to develop this code in an IDE or text editor, and copy it to the Field Calculator when you are done.

```
def fnFlow(pipeSize,pipeMaterial):
    if pipeSize == 4:
        return ""
    elif pipeSize == 6:
        return ""
    elif pipeSize == 8:
        return ""
    elif pipeSize == 10:
        return ""
    elif pipeSize == 12:
        return ""
```

6.   Move on to step 3 of the pseudo code, and add the nested if statement to determine material type. The if and elif statements are indented two spaces from the pipe size condition statement.

```
def fnFlow(pipeSize,pipeMaterial):
  if pipeSize == 4:
    if pipeMaterial[0] == "H":
      return ""
    elif pipeMaterial[0] == "D":
      return ""
    elif pipeMaterial[0] == "P":
      return ""
    elif pipeMaterial[0] == "R":
      return ""
    else:
      return ""
  elif pipeSize == 6:
    ....
```

Note the use of the slice function to get the first letter of each description in the if statement. This function keeps you from having to type the entire description, but make sure that when you use this function you are producing a unique value for each if statement. Although only one set of condition statements for pipe materials is shown, these statements need to occur for each pipe size.

7.   Finally, you must construct the calculation, and add it for each condition. Note: only part of the final script is shown in the graphic.

```
def fnFlow(pipeSize,pipeMaterial):
  if pipeSize == 4:
    if pipeMaterial[0] == "H":
      return 30 * 0.95
    elif pipeMaterial[0] == "D":
      return 30 * 0.82
    elif pipeMaterial[0] == "P":
      return 30 * 0.95
    elif pipeMaterial[0] == "R":
      return 30 * 0.88
    else:
      return 30 * 0.90
  elif pipeSize == 6:
    if pipeMaterial[0] == "H":
      return 70 * 0.95
    elif pipeMaterial[0] == "D":
      return 70 * 0.82
    elif pipeMaterial[0] == "P":
      return 70 * 0.95
    elif pipeMaterial[0] == "R":
      return 70 * 0.88
    else:
      return 70 * 0.90
  elif pipeSize == 8:
    if pipeMaterial[0] == "H":
      return 175 * 0.95
    elif pipeMaterial[0] == "D":
      return 175 * 0.82
    elif pipeMaterial[0] == "P":
      return 175 * 0.95
    elif pipeMaterial[0] == "R":
```

8. Once you have the entire script in the Field Calculator code box, click Save, and save to your MyExercises folder.

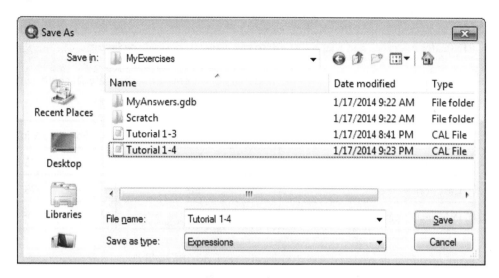

9. Click OK to run the code, and see the results calculated into the field. (Hint: make sure Python is still selected as the parser in the Field Calculator dialog box—it sometimes resets to the default of VB Script.)

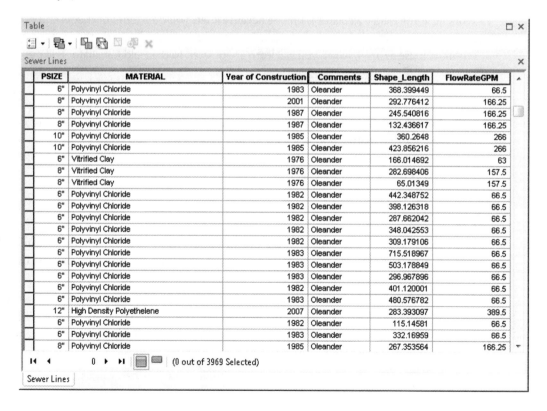

If your code is not running correctly, double-check the variable names, the indentations, the colons, and so on. For long, complex scripts like this one, you can do them in your IDE, which checks syntax as you go, and then copy and paste the final script to the Field Calculator code box.

Exercise 1-4

The city engineer has a similar project using the water line data. Open the map document Exercise 1-4 and look at the attribute table for the Distribution Laterals layer. Use the fields PSIZE, PTYPE (for material), and Shape_Length to calculate the desired use factor.

The formula is PSIZE * Shape_Length * Material Coefficient, using the material coefficients from the following table:

	For pipes 6 in. or less	For pipes 8 in. or larger
Asbestos concrete	80	95
Cast iron	60	75
Polyvinyl chloride	90	105

Write pseudo code to determine the process. Then use the Field Calculator to complete the process, and place the answer in the DiamLengthPressure field.

Tutorial 1-4 review

Working in the Field Calculator is unique because it lets you do a series of checks and calculations on a per-feature or per-row basis. Each feature is evaluated individually. To do this in stand-alone Python scripts, you use a *cursor*, which you will learn about later, so the Field Calculator is like an automatic cursor.

There are limitations to the Field Calculator, however. Although you can bring a number of field values from the current feature class into the script, you can only deal with one output field at a time—and for that matter, only one feature class or one table at a time. In a full Python script, you can control any number of field values, feature classes, or tables simultaneously and send output to other fields, other feature classes and tables, or even files outside ArcGIS.

Study questions

1. If you have a feature class with property values (Tot_Value) and a separate feature class with lot size (Acreage), could you write a script to calculate the value per acre using the Field Calculator? Write the script (if you think you can do it).

2. You need to calculate a property drainage coefficient based on lot size (Acreage), land use (Use_Code), an impervious area (Imperv_Area), and soil type (Dirt_Type). If all these fields are in the same feature class, can they all be used in the Field Calculator? Write the code to bring all these fields into the Field Calculator dialog box (if you think you can do it).

3. Name three things to watch for when nesting if statements.

Tutorial 1-5 Working with Python date formats

Dates are a complex item in any programming language, but Python makes using them easy. Basic formatting techniques are used to extract and manipulate data information.

Learning objectives

- Using Python date directives
- Working with date information
- Building complex Python objects

Preparation

Research the following topic in ArcGIS for Desktop Help:

- "Fundamentals of date fields"

Introduction

As you have seen in tutorials 1-3 and 1-4, complex calculations can be made in the Field Calculator using Python directives and if condition statements. One type of calculation that causes concern among programmers is performing date calculations because the fields that hold the dates are not standard fields, and they are not structured on a base 10 calculation like common numbers. A date field is a special type of field that can contain the day, month, year, and time of day in a variety of formats, which means that a standard Python variable is not able to contain the data. Instead, you must use a Python date object. In using the date object, make sure to specify the date components as they are brought into the object so that they can be retrieved as needed.

You also must pay attention to the time field. Merely subtracting the time values will not produce the desired results. The hours, minutes, and seconds must be calculated separately, and at the same time, you must also be aware of the scenario covering multiple days.

Scenario

The fire chief has provided you with the calls for service data for the past year and wants you to calculate the elapsed time between the dispatch time and the time that the vehicle arrived on the scene in decimal minutes. Any call that exceeds five minutes will need to be investigated. Although this sounds simple, it can be one of the more complex functions to perform.

Data

The calls for service data has the fields dispatched and arrived, which represent the time the vehicle left the station and the time it arrived on the scene.

SCRIPTING TECHNIQUES

Remember that variables typically hold a single value, but objects can hold multiple values. The list objects that you used earlier are a good example of objects with multiple values. The key to working with objects is to understand the format of what the object holds and how to access it. As the book progresses, you will see more examples of objects and how to research their structure.

Date objects can actually hold both date and time, or date only or time only, so particular care must be taken to assess the values in the object before deciding how to process them. Once this is done, identify each of the components using a format code called a *date directive.* The following list of directives will help you decide how to format the date object:

- %a = abbreviated weekday name
- %A = full weekday name
- %b = abbreviated month name
- %B = full month name
- %c = preferred date and time representation
- %C = century number (the year divided by 100, range 00 to 99)
- %d = day of the month (01 to 31)
- %D = same as %m/%d/%y
- %g = like %G, but without the century
- %G = four-digit year corresponding to the ISO week number (see %V)
- %h = same as %b
- %H = hour, using a 24-hour clock (00 to 23)
- %I = hour, using a 12-hour clock (01 to 12)
- %j = day of the year (001 to 366)
- %m = month (01 to 12)
- %M = minute
- %n = add a new line
- %p = either a.m. or p.m., according to the given time value
- %r = time in a.m. and p.m. notation
- %R = time in 24-hour notation
- %S = second
- %t = Tab character

- %T = current time, equal to %H:%M:%S
- %u = weekday as a number (1 to 7), where Monday = 1 (**Warning:** in the Sun Solaris operating system, Sunday = 1.)
- %U = week number of the current year, starting with the first Sunday as the first day of the first week
- %V = the ISO 8601 week number of the current year (01 to 53), where week one is the first week that has at least four days in the current year, and with Monday as the first day of the week
- %W = week number of the current year, starting with the first Monday as the first day of the first week
- %w = day of the week as a decimal, where Sunday = 0
- %x = preferred date representation without the time
- %X = preferred time representation without the date
- %y = year without a century (range 00 to 99)
- %Y = year including the century
- %Z or %z = time zone or name or abbreviation
- %% = a literal % character

You must use the Python DateTime module to correctly use the date object. This is a standard Python module that contains specialized functions and methods for working specifically with date information. The syntax is to add "from datetime import datetime" at the beginning of your code. This syntax brings in the date and time functions you will use in the calculations. For example, the date string "10/25/2012" would use the format string "%m/%d/%Y".

The DateTime module also includes special functions to perform math on dates. Simple subtraction of dates using the standard Python math functions could produce incorrect results. Imagine a call for service that started at 11:58 a.m. and ended four minutes later at 12:02 p.m. Performing a simple subtraction of these values would produce a negative number. The same would be true of a call that occurred just before midnight and ran into the next day. But once the values in the fields are put into date objects, subtracting them produces a time delta object. The function total_seconds() can be used to extract a value into a numeric variable, and dividing this by 60 converts the value to minutes.

Handling time is also tricky. Seconds and minutes are on a base 60 system, with hours being on a base 12 system, or a base 24 system for military time. You determine the base when you format the Python object. The directives handle almost any combination, but make sure you match them to the data carefully. This will ensure success with your code.

Work with the Python DateTime module

1. Open the map document Tutorial 1-5. In the table of contents, click the List By Source button, and open the table Calls_for_service_2012. Note the fields that you are working with.

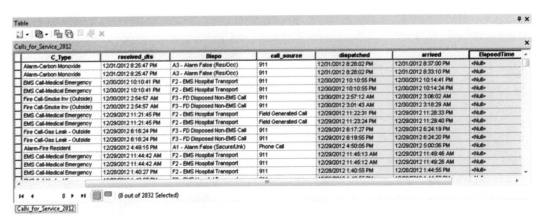

Notice the format of the date. The fields have the following structure:

- Month shown as two digits followed by a slash
- Day of the month shown as two digits followed by a slash
- Year shown as four digits followed by a space
- Hour in base 12 shown as one or two digits followed by a colon
- Minutes in base 60 shown as two digits followed by a colon
- Seconds in base 60 shown as two digits followed by a space
- AM and PM shown as two uppercase characters to determine morning or afternoon, respectively

The key to working with dates is to select the correct formatting string during the object assignment. Use the function .strptime() with the syntax strptime(date, format) where date is the date string you are bringing in from ArcMap, and format is the Python directive to identify each component of the date.

2. Write the general and detailed pseudo code necessary to calculate the elapsed time. Include the date string formatting directives to accept the data from the calls for service table. Use the preceding list in "Scripting techniques" to determine which directives to use. (As usual, try to write the complete pseudo code for each step before comparing your results with the graphics).

```
# Get the date fields from ArcMap.
#    1. Define a function to bring the fields "dispatched" and "arrived"
#       into the Field Calculator.
#       Create two variables to hold these values.
# Format the strings into Python date objects.
#    2. Select the correct directives for the date.
#       %m/%d/%Y %I:%M:%S %p
# Subtract the dates.
#    3. Subtract the formatted date objects to create a time delta object.
# Output the results in decimal minutes.
#    4. Use the total_seconds() function, and divide the results by 60.
#       Use round() to round the results to two decimal places.
```

3. Right-click the ElapsedTime field, and open the Field Calculator dialog box. Enter the code to set up the function, and bring in the necessary fields.

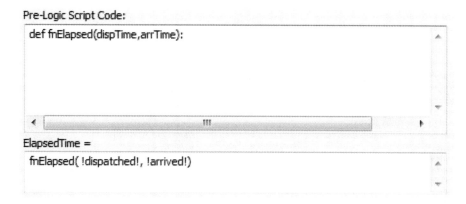

4. Add the code to format the date string. Import the Python date-handling method, and then create a Python object with the correct date format, as shown:

```
def fnElapsed(dispTime,arrTime):
    from datetime import datetime
    dateDispatch = datetime.strptime(dispTime,"%m/%e/%Y %I:%M:%S %p")
    dateArrive = datetime.strptime(arrTime,"%m/%e/%Y %I:%M:%S %p")
```

The date subtraction is next. The calculation must accommodate the possibility of a time that bridges the day or the morning to afternoon break. Subtracting the two date objects will result in a time delta object, which is designed to automatically accommodate the time changes. The time delta object holds the elapsed time in seconds and is retrieved using the total_seconds() function. When divided by 60, the result is the total elapsed time in decimal minutes.

5. Write your version of the code to perform the date subtraction and compare it to this:

```
def fnElapsed(dispTime,arrTime):
    from datetime import datetime
    dateDispatch = datetime.strptime(dispTime,"%m/%d/%Y %I:%M:%S %p")
    dateArrive = datetime.strptime(arrTime,"%m/%d/%Y %I:%M:%S %p")
    timeDiff = dateArrive - dateDispatch
    elapMin = timeDiff.total_seconds()/60
    return round(elapMin,2)
```

6. Add the return statement, and send the results back to the table. Click OK to see the results. By sorting the field in descending order, you can quickly see which calls exceeded the required response time.

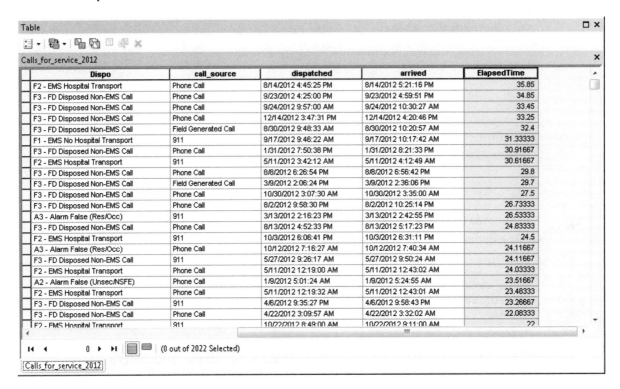

Dates can be problematic, but this example should help you understand how to do a variety of date calculations. The key is to use the correct directive when formatting the Python date object.

Exercise 1-5

Open the map document Exercise 1-5. Perform the same elapsed-time calculations on the Calls_for_ service_2010 data. Notice that the date and time data are in separate fields.

Write the pseudo code first to determine which directives are necessary to create the Python date objects.

Write the code in the Field Calculator, and store the results in the ElapsedTime field.

Tutorial 1-5 review

The examples in this tutorial cover how to properly format dates as input and place them into variables. Then these variables were used for calculations. The key to working with dates is to know the date format of the input value. Some date fields contain the full range of information, including date and time. Other fields may have only the date, and yet others may have only the time.

The date directives listed at the start of this tutorial also handle the formatting of output dates. For instance, you could format a date into a Python date object and print the corresponding day of the week using %u or the corresponding week of the year using %U or %W (depending on whether you start your week on Sunday or Monday).

Study questions

1. Research the date handlers in your Python reference book, and list the function to get today's date.

2. What format directives would you use on this date string: 31 12 2013 23:59:59 (New Year's Eve in London)?

3. What format directive would you use to give the full weekday name?

Chapter 2

Writing stand-alone Python scripts

Introduction

Much of the Python programming you create will be stand-alone Python scripts, which you may also bring into ArcGIS as script tools. The scripts will follow a standard format and are written using an integrated development environment (IDE), such as IDLE, which comes with Python; PythonWin, which is popular with many programmers; or PyScripter, which is gaining popularity and is used for the examples in this book. You may use any of these IDEs because the formatting of the scripts and the code you write will be the same, regardless of the IDE used. Please read the documentation of the IDE you choose to learn how to create, run, and save Python programs.

One of the most important things you can do in your code development is to write proper pseudo code. Pseudo code is a plain-text description and diagram of what your code should accomplish. You should note which modules you need to import, which tools you might need to use, the steps necessary to perform the desired operations, which variables you might need to set up and use, and any research notes about the Python code you write. This description is not intended to be code, but it may become documentation in the script for future reference. You wrote a little pseudo code in the first chapter of this book. The pseudo code for these next chapters should be well written and researched and contain detailed descriptions of your process. A simple outline of the pseudo code is found at the end of each tutorial. You should prepare your own pseudo code first, and then compare it with the outline provided.

This chapter reviews some of the basic structure of Python code, but you should use a good Python reference book for more detailed study. Each tutorial is designed to highlight particular techniques to interface your code with ArcGIS. As always, if you find a better or more efficient way to accomplish the same task, by all means, give it a try. Writing and troubleshooting the code is good practice.

Special introduction: Working with Python

Python is similar to other object-oriented programming languages because it deals with functions, classes, and objects. Esri has written a module for Python, named ArcPy, that adds special functions, classes, and objects that deal specifically with ArcGIS and its components. In addition to standard Python code elements, ArcPy also provides access to all the tools available in ArcToolbox. To be successful, learn more about these elements and how to find references to their use. These tool references are easy to find in the materials provided with ArcGIS. The easiest and most straightforward method to access these tools is to search for the tool in ArcGIS for Desktop Help. Another easy way to get a tool reference is to right-click the tool in ArcToolbox, and select Item Description. A third way is to search for the tool in the Search window, and click the link provided with the tool's name. Whichever way you choose, you will be provided with an explanation of the tool's function, the syntax for the tool's parameters, and code samples showing how to use the code in Python scripting.

The simplest code to program uses the ArcGIS tools. Any tool found in ArcToolbox can be brought into a Python script. The basic syntax for using these tools is to call the ArcPy module, add a period, add the tool name and then an underscore, followed by the alias of the toolbox where this tool resides, and then add a set of parentheses containing the parameters necessary to run the tool. For example:

Module.Tool_ToolboxAlias(param1, param2)

This syntax may seem difficult, but code samples are given in the tool reference for each tool. For example, the Buffer tool would be added like this:

```
arcpy.Buffer_analysis(in_features,out_feature_class,buffer_distance)
```

How do you know things like the toolbox alias and the tool parameters? Simple: look in the tool's reference. The Help for the Buffer tool includes the tool syntax, a description of the parameters, and a couple of sample scripts to show how the tool is used in Python. As good practice, you could copy and paste the tool syntax into your script as a comment so that as you write out the parameters, you are sure to put them in the correct order. Also, pay close attention to the capitalization. Remember, Python is case sensitive.

Here is the tool reference for Buffer (shown without the optional parameters):

Syntax

Buffer_analysis (in_features, out_feature_class, buffer_distance_or_field, {line_side}, {line_end_type}, {dissolve_option}, {dissolve_field})

Parameter	Explanation	Data Type
in_features	The input point, line, or polygon features to be buffered.	Feature Layer
out_feature_class	The feature class containing the output buffers.	Feature Class
buffer_distance_or_field	The distance around the input features that will be buffered. Distances can be provided as either a value representing a linear distance or as a field from the input features that contains the distance to buffer each feature.	Linear unit; Field
	If linear units are not specified or are entered as Unknown, the linear unit of the input features' spatial reference is used.	
	When specifying a distance in scripting, if the desired linear unit has two words, like Decimal Degrees, combine the two words into one (for example, '20 DecimalDegrees').	

Look up a few commonly used tools in ArcGIS for Desktop Help or some tools in ArcToolbox, such as creating a file geodatabase and performing a union, and note the syntax and code examples provided. If you are not familiar with the exact tool name, use the Search window to find it. Be careful because some tools that you may commonly access from the toolbar when working manually may have different names when used as ArcPy tools. Good examples of these tools include Select By Location on the menu, which becomes Select Layer By Location in ArcPy, and Export from the Table Options menu, which becomes Copy Rows in ArcPy.

In many cases, the tools accessed through ArcPy require the path to files or workspaces. The backslash (\) is a protected character in Python, so system paths that include a backslash must be handled in a certain way. There are three ways to handle these paths: (1) use a double backslash (\\); (2) substitute a forward slash (/); or (3) put a lowercase *r* at the beginning of the character string containing the path. The three examples shown in the graphic would all point to the path correctly in a Python script.

```
# Use a double backslash
featureClass1 = "C:\\EsriPress\\GISTPython\\Data\\City of Oleander.gdb\\City_Area"
# Substitute a forward slash
featureClass2 = "C:/EsriPress/GISTPython/Data/City of Oleander.gdb/City_Area"
# Place a lowercase r in front of the string
featureClass3 = r"C:\EsriPress\GISTPython\Data\City of Oleander.gdb\City_Area"
```

Another programming item provided by ArcPy is functions. Common Python functions include the string and math functions, such as str(x), which casts a numeric variable as a string variable, and round(x,n), which rounds a numeric variable to *n* digits. Many of these functions are used in chapter one. A list of all the ArcPy functions (too long to list here) is available in ArcGIS for Desktop Help. These functions may perform simple tasks, such as adding error or warning messages to your script or determining the current license level.

Functions that are more complex and useful are used more often. The first of these is the Exists() function. This function is used with any dataset object to see if it already exists. It is often necessary to make sure that the output of a tool does not exist before trying to create it, such as geodatabases, feature datasets, tables, and feature classes. If the output already exists, it may cause your script to malfunction, so it is useful to check for the existence of items first.

Another commonly used function is a list function, and ArcPy has several types of them. List functions are used to create list objects of a variety of items that you will use in your script. All the list functions are shown in the graphic:

ListDatasets
ListDataStoreItems
ListEnvironments
ListFeatureClasses
ListFields
ListFiles
ListIndexes
ListInstallations
ListPrinterNames
ListRasters
ListSpatialReferences
ListTables
ListToolboxes
ListTools
ListTransformations
ListUsers
ListVersions
ListWorkspaces

The result of a list function is a list object, which is used with looping structures to iterate through the list. For example, if you wanted to see if a field named ROW_Width was in a feature class, you would use the ListFields() function to create a list object of the fields, and then iterate through the list to see if that name were there. As you can see, many items can be included in lists.

One of the most important features of ArcPy is a script's capacity to get input from the user. Without this feature, your scripts would just do the same things over and over. Python includes a function named raw_input() to get user input, but ArcPy includes a special function to do this named GetParameterAsText(). This function is used when you intend to run your script as a script tool in a toolbox. Conversely, a function named SetParameterAsText() returns values from your script tool in the event that you are using the script tool in a model. More about the use of these functions is found in later tutorials as you begin to write more code and use other code samples.

The next concept to master in Python is working with objects. An object is used to hold multiple characteristics or parameters of an element you may be using in your code. Interestingly, objects can be created with simple definition statements in the same way as variables, or objects can be returned from functions.

For example, a feature class is defined as an object in ArcPy. With the simple code shown in the graphic, you can create a feature class object.

```
roadsFeatureClass = r"c:\workspace\Roads_Feature_class"
```

Although the code may look simple, a feature class is a complex object with a variety of characteristics and properties that you can access. For a feature class, these attributes may include the name, file type, path, and extension. The tricky part is accessing these attributes through your Python code, which is accomplished using the Describe() function. The Describe() function outputs a describe object with all the properties available for easy use. For instance, the code in the following graphic creates a feature class object that holds the street centerlines feature class. The Describe() function is used to create a describe object for the feature class, which contains properties of the feature class. The print statement shown will print the name of the feature class, and the .path method will retrieve the path to the folder where the feature class is stored.

```
import arcpy
roadsFeatureClass = r"C:\EsriPress\GISTPython\Data\City of Oleander.gdb\Street_Centerlines"
# roadsFeatureClass = arcpy.GetParameterAsText(0)  # Optional code to prompt for user input
descFC = arcpy.Describe(roadsFeatureClass)
print descFC.baseName
arcpy.env.scratchWorkspace = descFC.path
```

With the file location hard-coded to a specific geodatabase, this may not seem too important, but note the optional code shown that would ask the user to enter a feature class location. A describe object must be created from the feature class to extract the feature class properties. The code uses .baseName to get the name of the feature class and .path to get the data path, which then is used to set the scratch workspace.

The Describe() function can be used on just about any type of file and dataset brought into ArcGIS. The list in the graphic shows the variety of objects this function will act upon.

Describe Object Properties	Prj File Properties
ArcInfo Workstation Item Properties	Raster Band Properties
ArcInfo Workstation Table Properties	Raster Catalog Properties
CAD Drawing Dataset Properties	Raster Dataset Properties
CAD FeatureClass Properties	RecordSet and FeatureSet Properties
Cadastral Fabric Properties	RelationshipClass Properties
Coverage FeatureClass Properties	RepresentationClass Properties
Coverage Properties	Schematic Dataset Properties
Dataset Properties	Schematic Diagram Properties
dBASE Table Properties	Schematic Folder Properties
Editor Tracking Dataset Properties	SDC FeatureClass Properties
FeatureClass Properties	Shapefile FeatureClass Properties
File Properties	Table Properties
Folder Properties	TableView Properties
GDB FeatureClass Properties	Text File Properties
GDB Table Properties	Tin Properties
Geometric Network Properties	Tool Properties
LAS Dataset Properties	Toolbox Properties
Layer Properties	Topology Properties
Map Document Properties	VPF Coverage Properties
Mosaic Dataset Properties	VPF FeatureClass Properties
Network Analyst Layer Properties	VPF Table Properties
Network Dataset Properties	Workspace Properties

The first item listed (Describe Object Properties) has properties associated with whatever object you are describing, while the other properties are specific to their individual object type. The properties .baseName (the name of the item) and .path (the folder or workspace containing the item) are part of the common object properties of such things as .extension (the file extension if the item is a file) and .children (a list element of items in a workspace or feature dataset). One of the most useful methods is .dataType, which returns the item's data type, such as FeatureClass, Workspace, FeatureDataset, File, or LocalDatabase. The script in the following graphic can be used to print all the describe object properties of any item, and these properties could be used to check that the correct file type is being used, to store the workspace for data creation, or to capture an item's name.

```
#Import the ArcPy Module
import arcpy

# List all describe object properties
# Substitute any item into the Describe command for a list of properties
desc = arcpy.Describe("C:\ESRIPress\GISTPython\Data\City of Oleander.gdb\Planimetrics")

# Print some describe object properties
#
if hasattr(desc, "baseName"):
    print "Base Name: " + desc.baseName
if hasattr(desc, "catalogPath"):
    print "CatalogPath: " + desc.catalogPath
if hasattr(desc, "childrenExpanded"):
    print "Children Expanded: " + str(desc.childrenExpanded)
if hasattr(desc, "dataElementType"):
    print "Date Element Type: " + desc.dataElementType
if hasattr(desc, "dataType"):
    print "DataType:       " + desc.dataType
if hasattr(desc, "extension"):
    print "Extension: " + desc.extension
if hasattr(desc, "file"):
    print "File: " + desc.file
if hasattr(desc, "fullPropsRetrieved"):
    print "Full Properties Retrieved: " + str(desc.fullPropsRetrieved)
if hasattr(desc, "metadataRetrieved"):
    print "Metadata Retrieved: " + str(desc.metadataRetrieved)
if hasattr(desc, "path"):
    print "Path: " + desc.path
if hasattr(desc, "name"):
    print "Name:         " + desc.name


# Examine children and print their name and dataType
#
print "Children:"
for child in desc.children:
    print "\t%s = %s" % (child.name, child.dataType)
```

Each item type will also have a set of describe properties to access properties unique to the particular item, such as Text File Properties and Geometric Network Properties. For instance, to access additional properties of a feature class, you might look at the Dataset Properties, FeatureClass Properties, File Properties, GDB FeatureClass Properties, Layer Properties, and Table Properties to uncover a variety of different properties.

Classes are a more complex element in ArcPy and are used to create objects to hold tool parameters independent of feature classes and workspaces. The long list of these classes is found in ArcGIS for Desktop Help. These classes may be used to set up the parameters for items such as fields, spatial references, and extents before they are applied to existing geometry objects, or these classes may be used to store these parameters after they are extracted from geometry objects.

A common thread in using all these ArcPy features is ArcGIS for Desktop Help, which is beneficial for finding the syntax, usage, and code examples of these features. As stated before, it is good practice to copy and paste information from the Help files into your script as comments to remind you of how an element can be used.

By using combinations of functions, classes, objects, and tools, you will be able to write scripts that will do amazing things with your geographic data.

Tutorial 2-1 Creating describe objects

The ArcPy module provides access to all the geoprocessing tools for tasks you can perform in ArcGIS. These tasks can be done one at a time or used for batch processing with a loop routine.

Learning objectives

- Writing pseudo code
- Using the Describe() function
- Learning ArcPy module basics
- Using Python environment variables
- Using basic error handling

Preparation

Research the following topics in ArcGIS for Desktop Help:

- "A quick tour of Python"
- "Importing ArcPy"
- "Using environment settings in Python"

Introduction

This tutorial demonstrates some of the basic code structure to get your Python scripts to interface with ArcGIS. The main component is the use of the ArcPy module. A module is a special set of tools that you bring into Python to add functionality in extending the basic Python code. In tutorial 1-5, you used a Python module named DateTime, which adds special time handlers. ArcGIS users can import the Esri module ArcPy, which provides access to all the geoprocessing tools and many special features associated with spatial data. For more information about ArcPy, go to the ArcGIS for Desktop Help Contents tab, and navigate to Desktop > Geoprocessing > ArcPy > Introduction.

Scenario

You have received some geographic data in a folder, and you would like to get information about some of its elements. Try your new Python skills and learn more about the layout of a stand-alone Python script by writing a Python script to report the information. Here's the pseudo code for this script:

- Import Python modules.
- Set the workspace environment.
- Check the spatial reference for a feature class in the workspace.
- Print the results to the screen.

Data

In the Data folder of the online data that goes with this book is a file geodatabase named *Sample Data*, which contains the feature classes shown:

Name	Type
arbordaze2009tents	File Geodatabase Feature Class
complan	File Geodatabase Feature Class
libsprk	File Geodatabase Feature Class
ROW_And_Easements	File Geodatabase Feature Class
sprinklerunit	File Geodatabase Feature Class
ZIPCODES_poly	File Geodatabase Feature Class

SCRIPTING TECHNIQUES

This script and all the scripts that follow start with code to import the ArcPy module and the env class and then to set the workspace environment. Almost every script you write will perform these three tasks at the beginning. Importing the modules is not necessary if the script is run as a tool in ArcMap or ArcCatalog, but it does no harm to include this code. Other environment settings, such as adding the results to the current map document, also are used only if you are working from a script tool in a map document.

In addition, this script uses the Describe() function to get information about a feature class. The syntax is to make a variable equal to arcpy.Describe(feature class name), resulting in a describe object. This syntax can be used in a variety of ways, depending on the information you are extracting. The code shown creates a describe object and then uses that object to extract parameters of the feature class.

```
import arcpy
# Describe the feature class Elm_Fork_Addition, and make a describe object
streetDescObj = arcpy.Describe(r"C:\EsriPress\GISTPython\Data\OleanderOwnership.gdb\Elm_Fork_Addition")
# Get the properties of the feature class from the describe object
fcName = streetDescObj.baseName # Name of feature class
fcPath = streetDescObj.catalogPath # Path to the file
fcShape = streetDescObj.shapeType # Feature type as Point, Polyline, Polygon etc...
fcSpatIndex = streetDescObj.hasSpatialIndex # True/False for the existence of a spatial reference
```

Once the variables are set to the different parameters, they can be used in if statements and other constructs within the script.

Write loop routines

1. Start your IDE, and create a new file named Tutorial_2-1.py. **Add comments at the top of the code block to name and describe your program. Add the pseudo code as comments, which will help when you start entering the code.**

```
#-------------------------------------------------------------------------
# Name:          Tutorial 2-1.py
# Purpose:       Learn the basic format of Python code used in ArcGIS
#
# Author:        David W. Allen, GISP
#
# Created:       10/25/2013
# Copyright:     (c) David 2013
#
#-------------------------------------------------------------------------

#Import Python modules

#Set the workspace environment

#Check the spatial reference for a feature class in the workspace

#Print the results
```

Start by writing the code. The first step is to import the ArcPy module. If this script were to run in ArcGIS, this step would not be needed, but it does no harm to include it in case you want to run the script outside ArcGIS.

The standard format to add the ArcPy module is to type "import arcpy." Arcpy contains several functions, classes, and modules, which supply specialized tools. The env class provides access to the ArcGIS environment setting; the data access (arcpy.da) module provides access to various data access tools and functions; and the mapping (arcpy.mapping) module provides access to the settings and characteristics of map documents. These modules are imported as part of ArcPy, or they can be imported separately by using the format "from arcpy import env."

2. After the first line of pseudo code, add the statements import arcpy **and** from arcpy import env, **as shown. This code will import the modules necessary to access the ArcGIS tools and the file geodatabase.**

```
#Import Python modules
import arcpy
from arcpy import env

#Set the workspace environment

#Check the spatial reference for a feature class in the workspace

#Print the results
```

Notice that the env class was also loaded, which will help shorten the instructions you will need to write in your code to access the env properties. A few of the more commonly used properties are shown here, but the entire list can be found in ArcGIS for Desktop Help.

Property	Explanation	Data Type
addOutputsToMap (Read and Write)	Set whether tools' resulting output datasets should be added to the application display.	Boolean
referenceScale (Read and Write)	Tools that honor the Reference Scale environment will consider the graphical size and extent of symbolized features as they appear at the reference scale. Learn more about referenceScale	Double
scratchFolder (Read Only)	The Scratch Folder is the location of a folder you can use to write file-based data, such as shapefiles, text files, and layer files. It is a read-only environment that is managed by ArcGIS. Learn more about scratchFolder	String
scratchGDB (Read Only)	The scratch GDB is the location of a file geodatabase you can use to write temporary data. Learn more about scratchGDB	String
scratchWorkspace (Read and Write)	Tools that honor the Scratch Workspace environment setting use the specified location as the default workspace for output datasets. The Scratch Workspace is intended for output data you do not wish to maintain. Learn more about scratchWorkspace	String
workspace (Read and Write)	Tools that honor the Current Workspace environment setting use the workspace specified as the default location for geoprocessing tool inputs and outputs. Learn more about currentWorkspace	String

3. **Add code to set the workspace property to the location where you installed the student data, as shown:**

```
#Set the workspace environment
env.workspace = r"C:\EsriPress\GISTPython\Data\Sample Data.gdb"
# arcpy.env.workspace = r"C:\EsriPress\GISTPython\Data\Sample Data.gdb"
```

Note the second line in this graphic is how you would call the env class if you had not used the from-import statement earlier. Any of the env properties can be set in the same manner, as shown:

```
env.scratchWorkspace = r"C:\EsriPress\GISTPython\Data\City of Oleander.gdb"
env.addOutputsToMap = True
env.referenceScale = '25000'
```

Following the guide of the pseudo code, the next step is to find the spatial reference of a named feature class. Finding the spatial reference involves creating a Python object to contain the properties of the feature class. These objects are created the same way that variables are created.

4. Because you have already set the workspace, the Python script will know where to look for the feature class you name, so set a variable named fcName to ZIPCODES_poly, as shown:

```
#Check the spatial reference for a feature class in the workspace
fcName = "ZIPCODES_poly"
```

Once this variable is set, you can access the properties of the feature class using the Describe() function, including the name, feature type, path, file extension, and extent. The table shows a few of these properties, including the presence of a spatial reference.

Property	Explanation	Data Type
canVersion (Read Only)	Indicates whether the dataset can be versioned	Boolean
datasetType (Read Only)	Returns the type of dataset being described • Any • Container • Geo • FeatureDataset • FeatureClass • PlanarGraph • GeometricNetwork • Topology • Text • Table • RelationshipClass • RasterDataset • RasterBand • TIN • CadDrawing • RasterCatalog • Toolbox • Tool • NetworkDataset • Terrain • RepresentationClass • CadastralFabric • SchematicDataset • Locator	String
spatialReference (Read Only)	Returns the SpatialReference object for the dataset	SpatialReference

The syntax for describing a layer is:

```
arcpy.Describe(layername)
```

5. Write the code to create a describe object named fcNameProperties **that references the properties of fcName. Check your code against the code shown:**

```
#Check the spatial reference for a feature class in the workspace
fcName = "complan"
fcNameProperties = arcpy.Describe(fcName)
```

By adding the .spatialReference parameter at the end of the describe object, you will gain access to the many spatial reference properties of the layer. These properties include all the technical specifications of the spatial reference, but you are interested in the name. The syntax for getting the name of the spatial reference is:

```
print describeObject.spatialReference.name
```

6. Write the code to print the name of the spatial reference of fcNameProperties. **If your script matches the one shown, save and run it.**

```
#-----------------------------------------------------------------------
# Name:        Tutorial 2-1.py
# Purpose:     Learn the basic format of Python code used in ArcGIS
#
# Author:      David W. Allen, GISP
#
# Created:     10/25/2013
# Copyright:   (c) David 2013
#
#-----------------------------------------------------------------------

#Import Python modules
import arcpy, sys
from arcpy import env

#Set the workspace environment
env.workspace = r"C:\EsriPress\GISTPython\Data\Sample Data.gdb"
# arcpy.env.workspace = r"C:\EsriPress\GISTPython\Data\Sample Data.gdb"

#Check the spatial reference for a feature class in the workspace
fcName = "ZIPCODES_poly"
fcNameProperties = arcpy.Describe(fcName)

#Print the results
print fcNameProperties.spatialReference.name
```

The feature class has the spatial coordinate system as shown:

```
Python Interpreter

NAD_1983_StatePlane_Texas_North_Central_FIPS_4202_Feet
>>>
```

Your turn

Modify the code to examine the feature classes arbordaze2009tents and complan, and determine whether these classes have a spatial reference.

What would happen if you misspelled one of the layer names? The code would have some serious issues with variables that do not exist and a Describe() function that does not work.

7. Add some error handling to your code with the Python try and except statements. Basically, add try: at the top of your code and except: at the bottom. After the except statement, add a print statement to warn the user of an error. The basic syntax is:

```
try:
    #
    #
    # All your code goes here
    #
    #
except:
    print "Your warning message"
```

Note that the try-except routine requires that you indent all your code, in addition to indenting your warning message at the end.

8. As shown, add the try-except error handing to your script with the message "I couldn't find that feature class!" Then type an incorrect file name and run it.

```
#-------------------------------------------------------------
# Name:        Tutorial 2-1.py
# Purpose:     Learn the basic format of Python code used in ArcGIS
#
# Author:      David W. Allen, GISP
#
# Created:     10/25/2013
# Copyright:   (c) David 2013
#
#-------------------------------------------------------------
try:
    #Import Python modules
    import arcpy, sys
    from arcpy import env

    #Set the workspace environment
    env.workspace = r"C:\EsriPress\GISTPython\Data\Sample Data.gdb"
    # arcpy.env.workspace = r"C:\EsriPress\GISTPython\Data\Sample Data.gdb"

    #Check the spatial reference for ZIPCODES_poly
    fcName1 = "ZIPCODES_poly"
    fcName1Properties = arcpy.Describe(fcName1)

    #Check the spatial reference for arbordaze2009tents
    fcName2 = "arbordaze2009tents"
    fcName2Properties = arcpy.Describe(fcName2)

    #Check the spatial reference for complan
    fcName3 = "complan"
    fcName3Properties = arcpy.Describe(fcName3)

    #Print the results
    print fcName1Properties.spatialReference.name
    print fcName2Properties.spatialReference.name
    print fcName3Properties.spatialReference.name

except:
    print "I couldn't find that feature class!"
```

This example is simplistic error handling, and you will learn more about error handling later.

Exercise 2-1

Write the pseudo code and script to discover the shape type in the specified feature class name. (**Hint:** Search ArcGIS for Desktop Help for the shapeType property of the Describe() function, which will explain the syntax for using this function. Check the describe object properties for a property to get the file's base name and the feature class properties for a property to get the file's shape type.)

Tutorial 2-1 review

The format for this Python script is the basis for every Python script you write, with the documentation first, the import arcpy statement (and any other module you need) coming next, and then the code to perform the operation. You can make a template file of this script format for future use. This template will not only simplify your code writing in the future, but also standardize your code. The inclusion of comments is important to help document your pseudo code. These comments also are helpful if you or another programmer must alter this script at some point in the future. This alteration may be made weeks or years later, and relearning the process this code is performing may be difficult without these comments.

The use of the Describe() function is also an important process to learn. Many of the properties of your data can be accessed by describing them to an object. ArcGIS for Desktop Help is helpful in finding all these properties. It is important to note that there may be more than one describe group for any one item, so searching Help is a good way to find all the groups. For instance, you could describe a feature class to an object and return characteristics as shown under properties, including Describe Object Properties, Dataset Properties, FeatureClass Properties, and File Properties.

Study questions

1. Using ArcGIS for Desktop Help, how many ways can you find to access the describe properties of a table?
2. What is the ArcPy module, and why is it used for geoprocessing scripts? Give examples of the modules, classes, and functions contained in the ArcPy module.
3. How would you find the available properties in the arcpy.environment class?

Tutorial 2-2 Scripting geoprocessing tasks

Almost any geoprocessing task can be automated with Python. This automation can help increase productivity for full-time GIS users or allow complex tasks to be shared with casual GIS users.

Learning objectives

- Programming geoprocessing tasks
- Understanding tool syntax

Preparation

Research the following topics in ArcGIS for Desktop Help:

- "Using the Results window"
- "Union (Analysis)"
- "Select (Analysis)"

Introduction

The environment settings and describe objects you worked with in tutorial 2-1 demonstrate the characteristics of feature classes. Now you are ready to start using these feature classes in geoprocessing tasks.

All the geoprocessing tools in ArcGIS are available through ArcPy. Each tool description provides both simple and complex Python code examples so that you can get an idea of how to use the tool in your scripts. Simply copy the examples into your own scripts, and modify them to use different workspaces and datasets.

Run the tools in ArcMap to test your processes. Then copy a code snippet from the process results to use in your scripts. For the more adventurous, drop the tool into a model, configure it, and export the model to a Python script. If done tool by tool, you could systematically extract almost all the code needed for a new script.

Scenario

You are creating a new wall map for the city planner that she intends to hang in her office. The map should show the parcels color-coded by the zoning classification. The datasets you have are parcels that have neither the zoning codes nor the zoning districts data that overlaps the streets. The objective is to create a dataset to use in this map that will color-code each parcel by its zoning classification without including the streets.

To create this dataset, union the parcels and the zoning districts to create a new feature class, and then remove the segments that are in the street right-of-way.

Data

The map document for this tutorial has the property lines for reference, the parcel data, and the zoning districts data, which is already classified by zoning code. The workspace for this data is as follows:

C:\EsriPress\GISTPython\Data\City of Oleander.gdb

(Modify it if necessary to match the location where you installed the student data.)

Script geoprocessing tasks

1. Write the pseudo code for this script. A completed version of the pseudo code is shown at the end of this tutorial. Write your pseudo code, and then compare it with the pseudo code provided. If your pseudo code differs and seems viable, work the tutorial as written, and then go back and try it again with your unique pseudo code.

2. Open the map document Tutorial 2-2. Note that the zoning categories shown overlap the streets—that is the part you will remove.

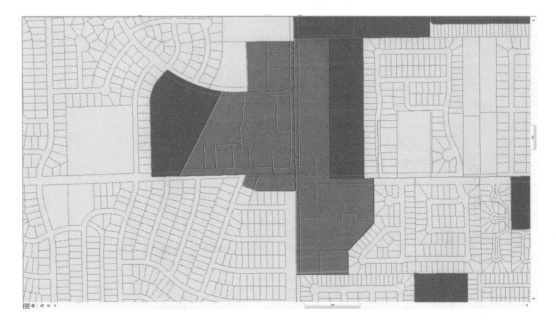

3. Start a new Python script in your IDE, and name it ZoneUnion.py. Add all the standard lines of code that you will need in your script, as shown:

```
#-------------------------------------------------------------------
# Name:        Tutorial 2-2
# Purpose:     Union two feature classes and delete streets
#
# Author:      David W. Allen, GISP
#
# Created:     10/25/2013
# Copyright:   (c) David 2013
#
#-------------------------------------------------------------------

try:
    # Import the modules
    import arcpy
    from arcpy import env

    # Set up the environment

    # ENTER CODE

    # Determine results

except:
    print "Message"
```

This template of code can be the basis for almost every script you write.

4. Add the code to set the workspace environment setting to the location of the parcels and zoning data. For help, refer to tutorial 2-1.

Next, union the parcels and the zoning districts layers. There are several ways to get a code snippet of this tool, so for this tutorial, use the geoprocessing Results window.

5. In ArcMap, open the Search window, and find the Union tool. Run this tool with the parameters shown:

6. When the tool is finished running, open the Geoprocessing > Results window from the main ArcMap toolbar. Expand the Current Session line, right-click Union, and click Copy As Python Snippet, as shown. Close the Results window.

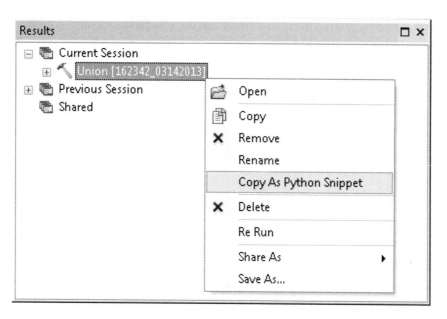

7. In your IDE, start a new indented line after the ENTER CODE comment line, right-click, and click Paste. In the example shown, a backslash was added to wrap the code onto a second line.

```
# ENTER CODE
arcpy.Union_analysis("Parcels #;ZoningDistricts #",\
"C:/EsriPress/GISTPython/MyExercises/MyAnswers.gdb/UnionTemp","ALL","#","GAPS")
```

You can see that the code for the Union tool starts with arcpy to access the ArcPy module, and then includes the tool name. Next are an underscore and the word *analysis*, which calls out the alias of the toolbox that contains this tool. A set of parentheses contains all the parameters for running this tool. The snippet will contain the raw path, so add a lowercase *r* at the beginning of the path to signal Python to interpret this string as a path and not as a string with special escape characters.

Notice that because you used a different file location for the output than the environment you set earlier, you will need to include this entire string when this file is used in other commands.

Also, be careful using this process of getting code snippets when you are writing a stand-alone script. The code snippet will use the layer alias from the table of contents, but when you run the script, it will look for the actual layer name. Any layer with an alias will need to be replaced in the code with the actual layer name.

8. **In ArcMap, open the attribute table of the UnionTemp layer. Right-click the FID_Parcels field and click Sort Ascending. Notice the records with a value of -1. Close the attribute table.**

The records with a value of -1 are the parts of the zoning data that fall in the street right-of-way. To remove these records, create a query to select all but these features, and then write the selected features to a new feature class. The Select tool in the Analysis toolbox will create a new feature class that contains only those features that meet a selection query.

9. **Use the Search window to find the Select tool. Run the tool, set the tool to output only the records from UnionTemp where FID_Parcels are <> -1, and store the records in a new feature class named** UnionFinal. **When your parameters match those shown in the graphic, click OK.**

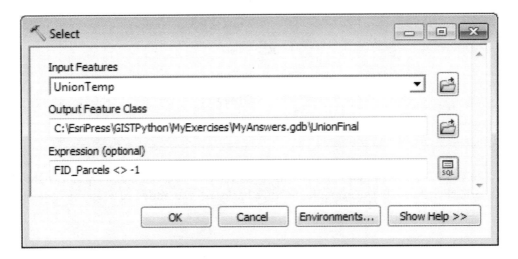

10. **Using the technique shown in step 6, copy the Python snippet for the Select tool from the geoprocessing Results window.**

11. **In your IDE, add a new indented line below the Union command, and paste the code snippet for the Selection tool. Add the full path for the UnionTemp layer in the command and the *r* at the beginning of the paths. Some additional comment lines are added, which you can delete.**

```
arcpy.Select_analysis(r"C:/EsriPress/GISTPython/MyExercises/MyAnswers.gdb/UnionTemp",\
"C:/EsriPress/GISTPython/MyExercises/MyAnswers.gdb/UnionFinal","FID_Parcels <> -1")
```

Note: Versions of ArcGIS prior to 10.2.1 may not format the query string in this command correctly. To accommodate this deviation, you must rework the query to use single or double quotation marks as shown:

```
arcpy.Select_analysis(r"C:/EsriPress/GISTPython/MyExercises/MyAnswers.gdb/UnionTemp",\
"C:/EsriPress/GISTPython/MyExercises/MyAnswers.gdb/UnionFinal",'"FID_Parcels" <> -1')
```

The output will be a feature class that shows zoning for each parcel but not in the street right-of-way.

12. Add a print statement after the "Determine results" comment, and change the error message at the end to something more appropriate for error handling, as shown:

```
try:
    # Import the modules
    import arcpy
    from arcpy import env

    # Set up the environment
    env.workspace = r"C:\EsriPress\GISTPython\Data\City of Oleander.gdb"

    # ENTER CODE
    arcpy.Union_analysis("Parcels #;ZoningDistricts #",\
    "C:/EsriPress/GISTPython/MyExercises/MyAnswers.gdb/UnionTemp","ALL","#","GAPS")

    arcpy.Select_analysis(r"C:/EsriPress/GISTPython/MyExercises/MyAnswers.gdb/UnionTemp",\
    "C:/EsriPress/GISTPython/MyExercises/MyAnswers.gdb/UnionFinal",'"FID_Parcels" <> -1')

    # Determine results
    print "New feature class has been created"

except:
    print "Process did not complete"
```

Next, delete the file UnionTemp from the geodatabase.

Your turn

Use the Search window to find a tool that will delete the feature class UnionTemp from the MyAnswers geodatabase. Configure the tool correctly, and add it to your script. Pay attention to the path names, and use the lowercase r if necessary to identify the path.

13. In ArcMap, use the Catalog window to delete the feature classes UnionTemp and UnionFinal that you created in testing. In your IDE, save your code, and run it, as shown:

```
#---------------------------------------------------------------
# Name:        ZoneUnion.py
# Purpose:     Union two feature classes and delete streets
#
# Author:      David W. Allen, GISP
#
# Created:     10/25/2013
# Copyright:   (c) David 2013
#
#---------------------------------------------------------------

try:
    # Import the modules
    import arcpy
    from arcpy import env

    # Set up the environment
    env.workspace = r"C:\EsriPress\GISTPython\Data\City of Oleander.gdb"

    # ENTER CODE
    arcpy.Union_analysis("Parcels #;ZoningDistricts #",\
    "C:/EsriPress/GISTPython/MyExercises/MyAnswers.gdb/UnionTemp","ALL","#","GAPS")

    arcpy.Select_analysis(r"C:/EsriPress/GISTPython/MyExercises/MyAnswers.gdb/UnionTemp",\
    "C:/EsriPress/GISTPython/MyExercises/MyAnswers.gdb/UnionFinal",'"FID_Parcels" <> -1')

    arcpy.Delete_management("C:/EsriPress/GISTPython/MyExercises/MyAnswers.gdb/UnionTemp","FeatureClass")

    # Determine results
    print "New feature class has been created"

except:
    print "Process did not complete"
```

When your script is finished running, a new feature class will be created that meets the city planner's requirements. Add the script to your map document, and import the same symbol schema from the ZoningDistricts layer.

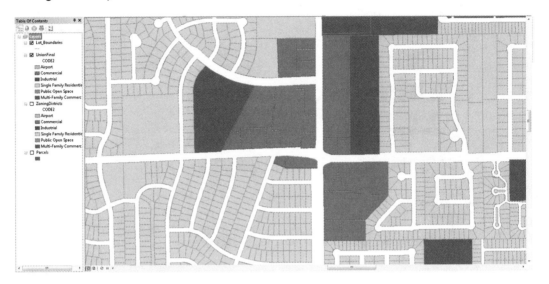

Here's the pseudo code for this project:

```
# Import the ArcPy module and env class
# Set up the working environment
# Perform the Union of the Parcel and Zoning District layers
# The parcel polygons that do not overlap the zoning polygons
# will be the street right-of-way and have an ID of -1
# Select out the parcels with an ID of -1
# Output a new layer with the results
# Symbolize to match the existing zoning layer
```

Exercise 2-2

The fire chief needs an updated map of the Fire Department's response zones, called *boxes*. The data layer FireBoxMap contains these zones. He would like the zones to have a gap between them so that they stand out on the printed maps. For aesthetics, you could symbolize the zones with a graduated color at the edges. The problem is, how is the feature for this type of display created?

Here is a description of the process from which you can write your pseudo code, research the tools necessary, and eventually write the script:

Convert the FireBoxMap polygons to a linear feature class, buffer them 50 feet, union the result with the FireBoxMap layer, and remove the features that represent the gap. Have your script delete any temporary files made in the process.

If done correctly, it should look like this:

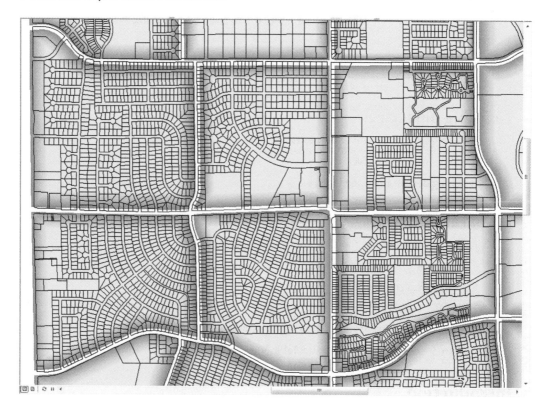

Tutorial 2-2 review

This tutorial uses some geoprocessing tools that you are probably familiar with from projects you may have done in ArcMap. When running the tools manually, you filled in the parameters in a pop-up dialog box and started the overlay process. The geoprocessing Results window provided a code snippet that you can use in your own scripts, which is convenient because all the tool's required and optional parameters are already set. Having this reference is also a good way to study the code and see how various paths and values are handled.

Study questions

1. This process of color-coding each parcel by zoning classification used the Union tool. Could you have used a different overlay tool? If so, what changes would need to be made in the code? Create a code snippet in the geoprocessing results for your code.

2. Could you have found all the polygons in the output of the Union tool that represented the right-of-way? Add code to make that layer, and store it in a new feature class.

3. The input layers for the Union tool must be polygons. Can you write the code to make sure that the layers are of the type polygon before the Union tool runs? (**Hint:** look at the describe properties.)

Tutorial 2-3 Coding for multiple geoprocessing tasks

Combining geoprocessing tasks with features, such as decision-making logic and feature cursors, allows programmers to make sophisticated scripts.

Learning objectives

- Using cursors and for statements
- Making a feature layer
- Using decision-making logic

Preparation

Research the following topics in ArcGIS for Desktop Help:

- "Accessing data using cursors"
- "Make feature layer (Data Management)"
- "Buffer (Analysis)"
- "Select Layer By Location (Data Management)"
- "An overview of the Layers and Table Views toolset"

Introduction

Running the geoprocessing tools in ArcMap and copying a code snippet is an effective way to get the tool syntax with all the parameters set, but there are still some issues that need to be fixed. That method also will not handle decision-making processes such as if-elif routines. In this tutorial, you will explore another method of developing the tool syntax by researching the script examples in ArcGIS for Desktop Help.

Each tool in the ArcGIS environment and each function and class in ArcPy have a well-documented Help page. This page provides a description of the tool and its parameters, and it typically includes two Python scripting examples—a simple case and a complex case. The simple example will probably make sense to you now, and as you develop your Python skills, the more advanced examples will as well. In fact, you may pick up some good coding techniques from the more complex examples, which you can also use in future scripts.

Scenario

A gas well drilling company has made an application to drill several wells in Oleander. To perform this drilling, the company must notify and get a signed lease document from all the property owners that are located over the planned drill paths. The city's Engineering division has asked that you generate a set of lists of property owners for each drill path. Before each path is drilled, the city will hold a public meeting with those homeowners and the drilling company to work out any issues that might exist. Because you do not know when each well will go online, you will create a separate mailing list for each well path and have them on hand for use when they are needed.

To create the list, buffer the well path and select the properties that intersect the buffer. Different well path lengths require different buffer widths—the longer the path, the wider the buffer. The distances are as follows:

- For well paths less than 3,000 feet, the buffer width is 75 feet.
- For well paths over 3,000 feet but less than 4,000 feet, the buffer width is 175 feet.
- For well paths 4,000 feet and longer, the buffer width is 300 feet.

Data

The data includes the well paths for nine proposed well projects. Also in the map document is the parcel data with a field named Prop_Add, which contains the property address necessary for the mailing list. Start with the most important step—writing the pseudo code.

SCRIPTING TECHNIQUES

Two new techniques are introduced in this tutorial. The first technique uses a cursor to access the rows in the feature class's attribute table or in a stand-alone table. Three types of cursors exist: search, which returns read-only values to the script; insert, which allows you to insert new rows into a table; and update, which allows you to change and delete rows in a table.

Start by defining a cursor object. This object uses one of the cursor commands from the data access module in ArcPy. The cursors have two required arguments, which are the table name and the fields from the table to use in the cursor. A single field or a list object that has many field names can be used in the cursor. The fields are indexed in the cursor in the order in which you list them. For example, index 0 would be the first field in the list, and index 1 would be the second field. The code shown in the graphic defines one of each type of cursor using either a single field or a list of fields.

```
exampleFC = r"C:\EsriPress\GISTPython\Data\OleanderOwnership.gdb\Elm_Fork_Addition"
# Define a search cursor with a single field
searchCur = arcpy.da.SearchCursor(exampleFC,["UseCode"])
# Since only one field is specified, index 0 is the UseCode field

# Define an insert cursor with two fields
insertCur = arcpy.da.InsertCursor(exampleFC,["Prop_Des_1","Prop_Des_2"])
# Field index 0 is Prop_Des_1, and field index 1 is Prop_Des_2

# Define an update cursor with a list of fields
fieldList = ["Prefix","StName","Suffix","SuffDir"]
updateCur = arcpy.da.UpdateCursor(exampleFC,fieldList)
# Field indexes will be 0, 1, 2, and 3 in the order of the fields in the list
```

As an option, the cursor could have a query statement so that only a subset of the rows is accessed.

Once the cursor is created, you can use it to move through the table one row at a time, always moving forward. The for statement has two parameters, the name of the object representing the current row and the cursor. Following these parameters, variables are added to hold the field values from the current row. The code shown in the graphic sets up a variable for each of the four values from the preceding update cursor example.

```
# Use a for statement to go through all the rows
for currentRow in updateCur:
    # Store the field values for the current row in variables
    stPrefix = currentRow[0]
    stName = currentRow[1]
    stSuffix = currentRow[2]
    stSuffDir = currentRow[3]
```

When all your processing is done, and before the script ends, delete the current row object and the cursor object, as shown in the following graphic. Deleting these objects will remove the file locks on the feature class or table you used.

```
# Delete the objects for the current row and the cursor
del currentRow
del updateCur
```

This step is not needed for a search cursor because that type of cursor does not lock the file being accessed.

The other new technique in this tutorial is the use of feature layers. The advantage of using a feature layer instead of a feature class is that the feature layer is a temporary copy of the data that exists only in memory. Changes can be made to the items in a feature layer and the data structure itself without affecting the source file. The changes will not persist after the script ends unless they are explicitly saved to the script. The MakeFeatureLayer tool also allows you to add a selection clause to the process that lets you work with a subset of the data. For instance, the vacant property of Oleander could be put in a separate layer file with the MakeFeatureLayer tool and an optional selection clause of "UseCode = 'VAC.'"

Code multiple geoprocessing tasks

1. **Write the pseudo code for this project. Include tool references and notations of parameters and conditions that you must set.**

Make sure that only one feature is selected, and that you have the correct feature length to set the buffer. A completed version of the pseudo code is included at the end of this tutorial that you can use for comparison to your own.

2.   **Start ArcMap, and open the map document Tutorial 2-3. Also, start your IDE, and create a new Python script named** WellNotification.py.

You can use the basic template of a Python script from tutorial 2-2, but in this tutorial, add some additional environment settings. In tutorial 2-2, code was added to the script to delete temporary feature classes so that they would not become a permanent part of your data. This time, use a scratch workspace to hold the temporary outputs rather than store them in your permanent workspace. To avoid an error, set the geoprocessing environment to allow existing data to be overwritten.

3.   **In your IDE, set the workspace to** C:\EsriPress\GISTPython\Data\City of Oleander.gdb\Well_Data. **Add the environment setting to overwrite the output of geoprocessing tasks, if it exists. Compare your code with the code shown:**

```
# Set up the environment
env.workspace = r"C:\EsriPress\GISTPython\Data\City of Oleander.gdb\Well_Data"
env.overwriteOutput = True
```

4.   **Create a variable to hold the name of the feature class that contains the first well path, as shown:**

```
# Create a variable with the name of the subject feature class
fcName = "BC_South_3H_Path"
```

Now that the feature class is known, determine the distance to use in the buffering. The script should check the Shape_Length value for each drill path and set up a condition statement to determine the buffer distance.

To get the field value, use a cursor, which allows you to go through each row in the table for the feature class and get the field values, one by one. For these datasets, each feature class has only one feature, so the first returned value can be used to determine the buffer path.

Of the several types of cursors, use a search cursor for this task. Research SearchCursor in ArcGIS for Desktop Help or any other ArcPy reference you have available. Next, set up the cursor to find the Shape_Length field, and then store that value in a variable.

5.   **Write the code to create a search cursor named** wellCursor. **Add the statements to store the value of Shape_Length in a variable named** drillLength, **as shown:**

```
# The buffer distance is dependent on the length of the drill path
# Check to see how long the path is
wellCursor = arcpy.da.SearchCursor(fcName,["Shape_Length"])
for row in wellCursor:
    drillLength = row[0]
```

Use the length of the feature to determine the buffer distance. The scenario at the start of this tutorial describes the conditions for each distance.

6. Set up an if-elif-else statement to determine the correct buffer distance, and create a variable to store it named wellBuffDist, as shown:

```
# < 3000 feet uses a buffer of 75'
# >= 3000 feet and < 4000 feet uses a buffer of 175'
# >= 4000 feet uses a buffer of 300'
if drillLength < 3000:
    wellBuffDist = 75
elif drillLength >= 3000 and drillLength < 4000:
    wellBuffDist = 175
else:
    wellBuffDist = 300
```

Next, buffer the input feature class by the determined well buffering distance. Store the output in a file named **SelectionBuffer,** and store it in a separate workspace, **C:\EsriPress\GISTPython\ MyExercises\Scratch\TemporaryStorage.gdb.**

7. Find the tool documentation for buffer, and use the examples shown to set up the correct buffer statement.

```
# Perform the buffering
# Buffer_analysis (in_features, out_feature_class, buffer_distance_or_field,
#     {line_side}, {line_end_type}, {dissolve_option}, {dissolve_field})
arcpy.Buffer_analysis(fcName, \
    r"C:\EsriPress\GISTPython\MyExercises\Scratch\TemporaryStorage.gdb\SelectionBuffer", \
    wellBuffDist)
```

With the buffer completed, move on to the selection process. Use the new buffer to select the parcels that intersect it. If you were doing this manually in ArcMap, you would use the Select By Location tool from the ArcMap Selection menu. A counterpart is available in ArcPy named Select Layer By Location, but the ArcPy selection tool will act only on a feature layer, not on a feature class. You must write code to make the input feature class a feature layer, and then add code to make the selection.

8. Research the Make Feature Layer and Select Layer By Location tools, and use the code samples to determine the code for this project, as shown in the graphic. (Hint: use the full path name for the Parcels input layer because it is not coming from the default workspace you set.)

```
# Use buffer to select the parcels
# Make a feature layer to temporarily hold the input data
arcpy.MakeFeatureLayer_management(r"C:\EsriPress\GISTPython\Data\City of Oleander.gdb\Parcels", \
    "Parcels_Lyr")

# Use the feature layer in the slection process
arcpy.SelectLayerByLocation_management("Parcels_lyr", "INTERSECT",\
    r"C:\EsriPress\GISTPython\MyExercises\Scratch\TemporaryStorage.gdb\SelectionBuffer")
```

Now that you have the features within the buffer selected, the last step is to write them out to a new table that can be used with mail-merge software. Note that this should not be a feature class. A search of the Help files produces two tools that look like they might work for this step: Table To Table and Copy Rows. Research these tools, and determine which one would work best.

9. Using your research, add the code to write the selected features to a new table formatted as the input feature class name with the word MailList appended at the end, as shown. When the code is completed, save the script.

```
# Copy the selected features to a new table
arcpy.CopyRows_management("Parcels_lyr", r"C:\EsriPress\GISTPython\Data\\" \
+ fcName + "_MailList.dbf")
```

Note the double backslash (\\) at the end of the folder string. Because this is pointing to a folder and not a file, Python requires the additional formatting character—otherwise, no backslash would appear between the completed file path and the file name.

10. To test the script, run it and select one of the drill path feature classes from the City of Oleander.gdb\ Well_Data feature dataset.

11. Add the new database table to your ArcMap document and open it. It contains all the fields from the Parcels layer, including the one you need, as shown:

OID	OBJECTID	Prop_Des_1	Prop_Des_2	Acreage	DU	PlatStatus	UseCode	PIDN	Prefix	
0	5148	BEAR CREEK BEND ADDITION	BLK A LOT 24	0.174968	1		1 A1	1899 A 24		W
1	5217	MIDWAY PK	BLK 18 LOT 10	0.996592	0		1 F1	25940 18 10	W	HA
2	6652	MIDWAY SQUARE ADDITION	BLK A LOT 18	0.163653	1		1 A5	25975 A 18		EF
3	6658	MIDWAY SQUARE ADDITION	BLK A LOT 17	0.154063	1		1 A5	25975 A 17		EF
4	7196	WOODCREEK	BLK A LOT 43	0.102241	1		1 A5	47485 A 43		RC
5	7204	HARWOOD COURTS	BLK E LOT 4	0.103758	1		1 A5	17402 E 4		BF
6	7481	BEAR CREEK BEND ADDITION	BLK A LOT 32	0.207402	1		1 A1	1899 A 32		W
7	7482	BEAR CREEK BEND ADDITION	BLK A LOT 33	0.142065	1		1 A1	1899 A 33		W
8	7483	BEAR CREEK BEND ADDITION	BLK A LOT 34	0.114024	1		1 A1	1899 A 34		W
9	7487	BEAR CREEK BEND ADDITION	BLK A LOT 35	0.13864	1		1 A1	1899 A 35		W
10	7496	HARWOOD CROSSING	BLK 1 LOT 3	4.31566	0		1 F1	17403 1 3	N	M
11	6513	MIDWAY SQUARE ADDITION	BLK B LOT 3	0.160455	1		1 A5	25975 B 3		BF
12	6520	MIDWAY SQUARE ADDITION	BLK B LOT 4	0.162595	1		1 A5	25975 B 4		EF
13	6741	ARBOR GLEN ADDITION	BLK A LOT 72	0.066816	1		1 A4	796C A 72		AI
14	6743	ARBOR GLEN ADDITION	BLK A LOT 81	0.092001	1		1 A4	796C A 81		AI
15	6437	MIDWAY SQUARE ADDITION	BLK A LOT 25	0.152216	1		1 A5	25975 A 25		IN
16	6631	MIDWAY SQUARE ADDITION	BLK A LOT 19	0.304003	1		1 A5	25975 A 19		EF
17	6903	WOODCREEK	BLK E LOT 32	0.20849	4		1 B4	47485 E 32	E	DE
18	6378	MIDWAY SQUARE ADDITION	BLK C LOT 3	0.161747	1		1 A5	25975 C 3		BF
19	6407	MIDWAY SQUARE ADDITION	BLK A LOT 24	0.16022	1		1 A5	25975 A 24		IN
20	6415	MIDWAY SQUARE ADDITION	BLK A LOT 23	0.167889	1		1 A5	25975 A 23		BF
21	6439	MIDWAY SQUARE ADDITION	BLK B LOT 1	0.18062	1		1 A5	25975 B 1		BF
22	6448	MIDWAY SQUARE ADDITION	BLK A LOT 22	0.151095	1		1 A5	25975 A 22		RF

Table — BC_South_3H_Path_MailList — 1 ▶ ▶| (0 out of 177 Selected)

You can supply this table to the city engineers, and they can use it when necessary to notify the property owners as each well is drilled.

12. Test your script on one or two of the other drill paths.

Here's the pseudo code for this task:

```
# Import the modules
# Set up the environment
# Create a variable with the name of the subject feature class
# The buffer distance is dependent on the length of the drill path
# Check to see how long the path is
# < 3000 feet uses a buffer of 75'
# >= 3000 feet and < 4000 feet uses a buffer of 175'
# >= 4000 feet uses a buffer of 300'
# Perform the buffering
# Use buffer to select the parcels
# Make a feature layer to temporarily hold the input data
# Use the feature layer in the selection process
# Copy the selected features to a new table
```

Exercise 2-3

The city manager has been asked by the city council to determine whether there are any areas of town that do not have adequate nighttime illumination from street lights. He in turn has passed this task on to you. Your results will be used to determine where street lights need to be added or whether any existing lights should be moved in the long term. You must write a script to do this job because the locations will be changing over the next two years, and you will be running this analysis frequently.

The data in the map document Exercise 2-3 includes the current street light locations. There are four different types of lights, and each type requires a different buffer distance. The codes are stored in a field named Type and are listed here with their appropriate buffer distances:

MV (mercury vapor) = 125 ft

MVH (mercury vapor—high pressure) = 160 ft

SV (sodium vapor) = 100 ft

SVH (sodium vapor—high pressure) = 200 ft

The script you write will need to consider each feature and generate a buffer according to the light type. The final results can be overlaid on the city layer to reveal places that need more lighting. Creating a new feature dataset to hold the output files would be helpful.

Write the pseudo code and the script necessary to perform this task. (**Hint:** Try adding a field to the feature class to store the buffer distance, and then go through each feature with a cursor and determine which buffer distance should be used. When all the distances are set, run the buffer tool using the new attribute as the buffer distance.)

Tutorial 2-3 review

When used in a script, the tool parameters must be set in the code. These parameters can be a hard-coded value or even a variable. The required and optional parameters can be found by looking up the tool in ArcGIS for Desktop Help. The Help files also include sample scripts showing how the tools can be used in your own scripts. Note that some of the parameters may be optional and can either be included, left out completely, or skipped by providing the value of "" as a placeholder so that other optional parameters can be accessed further along in the tool's usage syntax.

This tutorial also uses cursors to access the data feature by feature. Cursors can be used to access the data or to change or add data. The syntax is to create the cursor object, which holds all the values of all the features in one place. The for statement can then be used to step through the features or rows one at a time for processing. Closing the cursor at the end of its use is important so that the features are not locked to further access.

The use of a feature layer is important in these situations because it provides a way to make selections and process the data without putting the source data in jeopardy. These types of layers can be created for tables, called *table view*, and you will see later how a query can extract a subset of the data. For this tutorial, the feature layer was required because the selection processes only work either in a map document that is referenced from the table of contents or in a feature layer or table view. These processes will not work on layers referenced by their source path.

Study questions

1. Name a few other geoprocessing tools, and look up these tools' syntax and sample scripts in ArcGIS for Desktop Help.

2. When is it useful to work with a copy of the data in a feature layer or table view rather than access the source data directly?

3. Look up the syntax for cursors in ArcGIS for Desktop Help, and show the different types.

Tutorial 2-4 Using while statements

The iteration of a script can be controlled with a while statement, which will cause the script to continue running until a condition is met. The number of iterations will change depending on the condition.

Learning objectives

- Using cursors with while statements
- Iterating through features

Preparation

Research the following topics in ArcGIS for Desktop Help:

- "Cursor (arcpy)"
- "Copy Features (Data Management)"

Introduction

In tutorial 2-3, you learned the simple use of a cursor. A cursor can be used to go through an entire dataset feature by feature or row by row and to perform an individual task on each item it finds. By default, geoprocessing tasks act only on the selected feature, and a cursor has the effect of making the current cursor item the only selected feature.

The cursor you used in tutorial 2-3 retrieved the value of a single field to be used in a condition statement. Cursors can access any field in the input dataset, which is specified in the cursor setup. The for statement is used with the cursor to handle the looping through the dataset. The for statement and all the code included in its indent level is run for each item in the cursor. Mastering the use of cursors is important in writing a script that can handle a variety of tasks.

In addition to a cursor, this tutorial uses another control statement, the while statement. The programmer sets up a condition, and the while statement will continue running the script or portion of a script until the condition is met. Once the condition is met, the rest of the code runs. The while statement is often used to process subsets of the data or to monitor the number of times a process is performed. One thing to be careful about, however, is that if no scenario exists in which the condition of the while statement equates to False, the while statement will never end.

Scenario

The library in Oleander bought a bookmobile at the Fort Worth Police Auction and is investigating the best way to put it in service. The library has decided that the bookmobile will make periodic stops at apartment complexes because of the density of people it can reach with one stop. The question is how to implement bookmobile stops for single-family residential customers.

The head librarian has an idea that he wants to investigate. He has selected six places in the city where they would be able to set up stops. The librarian wants to serve about 200 households at each stop, but he is not sure how far those 200 households would have to travel to get to the bookmobile. The process is complicated, so you will want to build a script to test locations. Also, the librarian will be selecting more locations for further research based on your initial results.

The process is to go through each of the six sites, one by one, and do some analysis on the household count. For each site, select all the parcels within 150 feet, and add up the number of dwelling units. If the number exceeds 200, stop the selection process. If the number does not exceed 200, select all the parcels within another distance of the currently selected set, and add up the households again. If the number does not exceed 200, repeat the selection until it does. Output the selected set to a new feature class, and move on to the next site.

Data

The map document for this tutorial includes a file named BookmobileLocations that contains the six sites the librarian chose. The field Marker contains the site name. Also included are the parcels with a field named DU, which has the count of dwelling units.

SCRIPTING TECHNIQUES

This script must do two things: iterate through a set of features, and perform a select operation at each iteration. A good practice is to write and test the code for the first process, and then add the second process. This practice simplifies the troubleshooting process because you will be limiting the number of errors that can occur. Set up and test a cursor to access each of the features in the input feature class. Once this cursor is running, add the second process.

This second process includes another type of iterator called a *while statement*. As you know, a cursor will go through the records one by one and stop when all the records have been accessed. With a while statement, you will define a condition and write code to process the data. This process repeats as long as the condition you set evaluates to True. As with the other types of condition statements and iterators, the processes associated with the while statement hold at an indent level to identify where the process stops.

This tutorial requires you to imbed one type of iterator inside another, which can be tricky. Make sure that each iterator has a condition that will cause it to complete. The cursor is easy because it completes when the last record is accessed, but the while statement is not as straightforward. If you were to unknowingly set a condition that could never be false, the while statement would never end. For this reason, it is sometimes advisable to add a counter inside the while statement to limit the number of possible iterations to a number known to be past the script's reasonable operation. For example, if you feel that the script can complete its tasks in 15 iterations or less for all cases, then stop the script once it reaches 20 iterations. Stopping the script involves imbedding a count variable inside the while statements and using an if statement to end the script if the count exceeds the predetermined maximum. An example of this action is given later, and you can add this as an option.

Use while statements

1. **Open the map document Tutorial 2-4.**

2. **Write the pseudo code for this project. (**Hint:** you will need a cursor to go through the six sites and a while statement to keep adding more parcels until the dwelling unit count exceeds 200.) Completed pseudo code is shown at the end of this tutorial.**

3. **Create a new Python script in your IDE, and name it** BookmobileAnalysis.py**. You can use the familiar template lines to start your script.**

4. Import the appropriate modules, and set the workspace environment to C:\EsriPress\ GISTPython\Data\City of Oleander.gdb, **as shown:**

```
try:
    # Import the modules
    import arcpy
    from arcpy import env

    # Set up the environment
    env.workspace = r"C:\EsriPress\GISTPython\Data\City of Oleander.gdb\\"
    env.overwriteOutput = True
```

Start by writing a cursor to access each of the individual sites in the BookmobileLocations feature class. Get the code for the cursor working first, and add the while statement later. Writing the code in this order will not only be easier to understand, but will also be easier to troubleshoot.

As a general housekeeping rule, make a feature layer for each of the feature classes you will be working with. In addition, the feature layer for parcels should include only the single-family households, which you can find by adding an optional selection clause (DU = 1). A feature layer is only a temporary copy of the data, so any changes made to it are also temporary unless specifically saved to the script.

5. **Make two feature layers named** Parcels_lyr **and** Locations_lyr **from the Parcels and BookmobileLocations layers, as shown. Add a query to the Parcels_lyr statement to select only the single-family units. Check the tool reference if you are unsure of the query syntax.**

```
    # Set up cursor for the bookmobile sites
    arcpy.MakeFeatureLayer_management("Parcels","Parcels_lyr",'"DU"= 1')
    arcpy.MakeFeatureLayer_management("BookmobileLocations","Locations_lyr")
```

Next, create a search cursor to go through the Locations_lyr file, which contains the setup sites for the bookmobile.

6. **Create a new search cursor named** siteCursor. **Have the cursor bring in the field Marker, which contains the unique site name for each location, as shown:**

```
    siteCursor = arcpy.da.SearchCursor("Locations_lyr","Marker")
```

This particular layer has six features, so you know that the siteCursor object will contain the six values of the field Marker. The for statement will reference each row in the attribute table, one at a time, and allow you to perform geoprocessing tasks for each value. The syntax is:

for {variable name to track the current row} in {cursor name}:

Note that the statement ends with a colon. The colon signals the start of an indent level, and all the code that maintains that level will be run once for each row. The end of the for statement is signified by going back one indent level.

7. Construct a for statement using the siteCursor object, as shown:

```
siteCursor = arcpy.da.SearchCursor("Locations_lyr","Marker")
for row in siteCursor:
```

The variable named row will now hold the value of the field Marker. It is an indexed list variable, so you must add an index number at the end of the variable to reference which field value to use. In this case, only one field was brought into the cursor, so row[0] will return that value. If other fields were brought into the cursor, they would be referenced in the order listed in the cursor statement, and the variable row[x] would hold their index number.

8. Create a new variable named siteName that holds the value of Marker at each iteration, as shown:

```
siteCursor = arcpy.da.SearchCursor("Locations_lyr","Marker")
for row in siteCursor:
    siteName = row[0]
```

The code from steps 6 and 7 is shown so that you can note the indent level. Maintain this level until the cursor routine is finished.

In each iteration through the cursor, select the location feature from Locations_lyr, and use that location to select all the parcels within 150 feet.

9. Research the various select tools available in ArcMap, and note the syntax of the code. In this example, the Select_analysis tool is used. Store the selected feature in your scratch folder as SiteTemp (MyExercises\Scratch\Temporary Storage.gdb\SiteTemp). Be careful to correctly format the where clause parameter in this tool's syntax. Write your code, and check it against the code shown:

```
# Select parcels within 150 feet of the new selection
arcpy.Select_analysis("Locations_lyr",\
    r"C:\EsriPress\GISTPython\MyExercises\Scratch\Temporary Storage.gdb\SiteTemp",\
    '"Marker" = \'' + siteName + "\'")
```

The formatting of the where clause is tricky. As an example, the resulting statement for the first site should be as follows:

"Marker" = 'Site 1'

Decode the sequence of characters used in the selection statement. Use your Python references for help.

After selecting the site feature, next select all the parcels within 150 feet. You have used this tool before and should be familiar with its syntax.

10. Add a Select Layer By Location tool, and configure it to use the selected feature from step 9 and a 150-foot search distance, as shown:

```
arcpy.SelectLayerByLocation_management("Parcels_lyr","WITHIN_A_DIST.
    r"C:\EsriPress\GISTPython\MyExercises\Scratch\Temporary Storage.gdb
    "150","NEW_SELECTION")
```

A condition statement called a *while loop* will be placed in this part of the code, but for now, finish the processing to be performed in the cursor. Export the selected features to a new feature class, and name it using the value from the field Marker.

11. **Research a tool to export features to a new feature class in your MyAnswers geodatabase. Find the proper syntax, and write your code before referencing the code shown:**

```
# Export the selected features to a new feature class
# Use the current site name as the file name
arcpy.CopyFeatures_management("Parcels_lyr", r"C:\EsriPress\GISTPython\MyExercises\MyAnswers.gdb\\" \
+ siteName.replace(" ","_"))

print siteName + " Output OK!"
# Move to the next site and repeat
```

There are two things to note here. First, the site names have a space in them, which is not allowed in a feature class name. You can use the Python string method .replace() to change the space to an underscore. Second, a print statement was added to print the site name and to confirm that the steps were completed successfully. If no print statement was added, you would get no indication of the script's status.

Next, test the code, and see if the cursor is working as expected.

12. **Save your script and run it.**

The script should create new feature classes in the MyAnswers geodatabase, as shown:

Put some of the files into your map document and see if they meet your expectations.

Now move on to the while statement.

A while statement sets a condition and continues looping through its processes until the statement is false. In this project, the count of selected features must be less than 200. Check your Python resource books or ArcGIS for Desktop Help for the syntax and usage of while statements. Basically, set up a variable to hold the count of parcels, and then start the while statement with the condition that the variable not exceed 200.

13. **In the section of the code for the while statement, add a variable to get the count of features selected in the parcels layer. Then set up the while statement using this variable, as shown:**

```
# Start a while statement until number of dwelling units exceeds 200
parcelCount = int(arcpy.GetCount_management("Parcels_lyr").getOutput(0))
print parcelCount
while parcelCount < 200:
    # All statements at this indent level are part of the while loop
```

Note that the while statement ends with a colon and starts a new indent level. Everything that is indented at this level runs within the while statement. To end the statement, drop back one indent level.

The variable that is controlling the while loop (parcelCount) is called a *sentry variable* because it watches over the loop and provides the value that will eventually cause it to end. This variable was initialized with an ArcPy function named GetCount(), which is useful in a lot of Python scripts. Note that because parcelCount is a Python object and not a simple variable, you must add .getOutput(0) at the end of the object to reference the value's index number.

When you use the sentry variable in the while statement, you must make sure that the value can be true on the first run. Otherwise, the loop will never run. The second thing to watch for is that the sentry variable is updated somewhere in the loop and that a condition eventually exists where the condition of the while statement is false. Otherwise, the loop will never end. Some programmers will nest a count variable in the while loop and cause the loop to end after a certain number of iterations are completed.

If the count of parcels is less than 200, make another selection to add all the parcels within a distance of the currently selected parcels to the selection set. This process will use Select Layer By Location with a distance of 150 feet—check the tool Help to determine how to set the overlap type to select features within a certain distance and how to set the selection type to add features to the current set.

14. **Add the selection statement, making sure to select the parcels, and add the parcels to the currently selected features. Copy the parcelCount statement to get a new count, as shown:**

```
# Add to the selected set all property within 150 feet and redo count
# SelectLayerByLocation_management (in_layer, overlap_type, select_features,
#                                  search_distance, selection_type)
arcpy.SelectLayerByLocation_management("Parcels_lyr","WITHIN_A_DISTANCE",\
"Parcels_lyr","150","ADD_TO_SELECTION")

parcelCount = int(arcpy.GetCount_management("Parcels_lyr").getOutput(0))
print parcelCount
# Exit the while statement when count exceeds 200
```

The selection will increase the number of currently selected features, so the value of the parcelCount variable will also increase. This condition will eventually cause the while loop to end.

As a bonus, add a count variable, and limit the while loop to eight iterations. The code would look something like this:

```
# This count routine will repeat eight times
# Initialize the count variable outside the while loop
myCount = 1
# Set the starting condition as True
theCondition = True

while theCondition == True:
    # Inside the while loop check the value
    if myCount == 8:
        # Set the condition to False
        theCondition = False
    else:
        print "The current count is " + str(myCount)
        myCount = myCount + 1

print "All done! The final count equals " + str(myCount) + "."
# Script is saved in \GISTPython\Data for reference
```

The code is now finished; save and run it. Drag a few of the output layers to your map document to see the results, as shown:

Here's the pseudo code for this project:

```
# Set up cursor for the bookmobile sites
# Select parcels within 150 feet of the new selection
# Start a while statement until number of dwelling units exceeds 200
# Add to the selected set all property within 150 feet and redo count
# Exit the while statement when count exceeds 200
# Export the selected features to a new feature class
# Use the value of the field Marker as the file name
# Move to the next site and repeat
```

Exercise 2-4

A similar project has cropped up in the Public Works Department. The City of Oleander runs its own public water supply system and must maintain a state certification with annual inspections. The water system has water sampling stations where samples can be drawn and sent to the state lab for testing. Oleander wants to move up to a level 1 certification, which will require analysis of the pipes immediately adjacent to the sampling stations.

Open the map document Exercise 2-4.

Your task is to write a Python script that will select the 10 adjacent pipes around each water sampling station and output these to a new file using the description as the name. Be careful because some of the sampling stations are closed and should not be included in the analysis. (**Hint:** use the selection type dealing with common boundaries between the features rather than a set distance to select adjacent pipes.)

Tutorial 2-4 review

Nesting looping statements is a way to perform multiple processes on your data. In this case, you nested a while statement inside the cursor. Each feature was selected one at a time, and a complex process was performed.

The while statement included a condition for the looping process. As long as the condition of the while statement remains true, the loop will continue. Make sure that at some point the condition equates to false for the loop to end, or your script might never complete its processing. As a fail-safe maneuver to prevent an endless loop, a while statement with a count variable adds additional control to the number of times the processes loop.

Study questions

1. Give an example of when you might use a while statement in your data processing.
2. Is there a limit to the number of loops that can be nested in each other?
3. Besides a counter, what other control(s) could have been included to make sure it was not an endless loop? (**Hint:** use the count of the features.)

Tutorial 2-5 Using lists and for statements

A list object can be used to store a list of such items as features, files, and tables. This list object can be used in a for statement to iterate through the list. This technique provides a loop that iterates a finite number of times—once for each item in the list.

Learning objectives

- Creating and using lists
- Creating and calculating fields

Preparation

Research the following topics in ArcGIS for Desktop Help:

- "Create lists of data"
- "ListFeatureClasses (arcpy)"
- "Add Field (Data Management)"
- "Calculate Field (Data Management)"

Introduction

In tutorials 2-3 and 2-4, you learned how to iterate through features in a feature class using a cursor and a for statement. You can also use the for statement to iterate through other items by placing them in a list object. A list object is a special type of Python variable that can hold a list of values, and these values are indexed just as you have seen with the Python text-handling functions.

List objects can be defined simply by providing the values separated by commas, but ArcPy has several special functions to create lists from elements commonly used in geoprocessing tasks. These functions include listing fields, layers, and raster files. There are also functions to list map layout components, such as data frames and layout elements. The output of these tools is, in fact, tuples, which are a special type of Python list in which the values contained in the list cannot be changed. Because the use of tuples in ArcGIS is almost identical to how you would use lists, the references here will continue to treat them as standard Python list objects.

Once these ArcPy list objects are created, you can use them with a for statement to iterate through the list contained in the object. Each iteration of the for statement will expose an object that can be used for feature classes that need processing, layout elements that need updating, fields that need changing, and any number of other tasks.

Scenario

The Fire Department needs a count of all the single-family homes and multifamily structures in each of the 44 fire response zones. The secretary in that department is trying to complete this task with a felt-tipped pen and an aerial photo. You see that she has already missed a few structures, misidentified a few, and is spending an inordinate amount of time on this project. You are going to write a script that will be faster and more accurate. The Fire Department's geodatabase contains a polygon feature class for each response zone, or box. You will need to add two fields to each box: one for the single-family count and one for the multifamily count. The secretary has already added the fields to a portion of the feature classes, but the rest of the classes do not have these fields.

Data

The FireDepartment geodatabase contains 44 feature classes representing the box zones. Note that the name structure for these feature classes is the phrase "FireBoxMap" followed by the box ID number. In addition, there is a polygon feature class with the building footprints. Each building has a use code in a field named UseCode (1 = single family, 2 = multi-family).

SCRIPTING TECHNIQUES

Because there are 44 different feature classes to be used in this project, the best course of action is to make a list of the feature classes and iterate through the list. A list object can be created using the ListFeatureClasses tool, which can be set to include only files that meet a specified criterion. Once the list object is created, a for statement can be used to iterate through the list. Use a combination list and for routine for feature classes, although this type of iteration can be done with any of the list tools shown at the start of this chapter.

The second technique introduced here is the use of the Calculate Field tool. This tool is used to write values into fields of the attribute table. Any number or text string can be created and stored in a variable, and then the variable can be written to the specified field. The unique aspect of this technique is that the fields are written one by one rather than as a batch, so individual steps that may not be appropriate for the entire dataset can be calculated here.

Use lists and for statements

1. Open the map document Tutorial 2-5, and note the BldgFootprints feature class. Open the attribute table, and make note of the field containing the use code (UseCode), as shown:

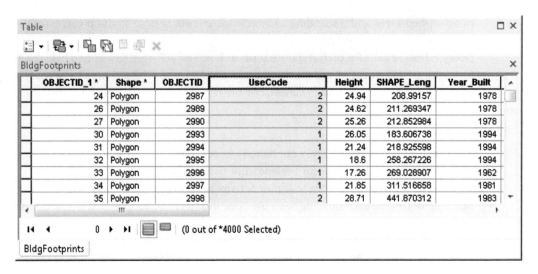

2. In the Catalog window, note the feature classes representing the 44 boxes, as shown:

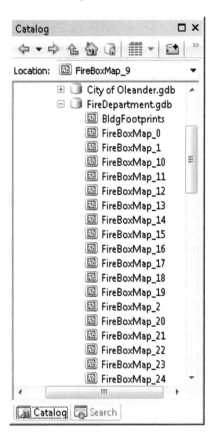

Develop a plan of action for tackling this project.

3. Write the pseudo code for this task, which will involve getting a list of all the layers and adding fields among other tasks. Completed pseudo code for this task is found at the end of this tutorial.

4. Start a new Python script named DwellingCount.py. Add the comments and template components as in the previous scripts, as shown:

```
#-------------------------------------------------------------------
# Name:        Tutorial 2-5
#              Dwelling count
# Purpose:     Count the buildings in each box zone.
#
# Author:      David W. Allen, GISP
#
# Created:     10/25/2013
# Copyright:   (c) David 2013
#
#-------------------------------------------------------------------

try:
    # Import the modules
    import arcpy
    from arcpy import env

    # Set up the environment

    # ENTER CODE

    # Determine results

except arcpy.ExecuteError:
    print  arcpy.GetMessages(2)
except:
    print "Process did not complete."
```

Notice that a different type of error handling has been added in the except line. The special arcpy.ExecuteError object will trap any errors raised by ArcPy statements. The second except statement will trap any other errors. Check the ArcPy documentation for more information on this object.

Start by making a list object of the feature classes in the FireDepartment geodatabase.

5.   Search ArcGIS for Desktop Help to locate an ArcPy function that can be used to get a list of feature classes. Several are shown in the table. Select the one you think is most appropriate, and open and read the tool reference.

ListFields(dataset, wild_card, field_type)	Returns a list of fields found in the input value
ListIndexes(dataset, wild_card)	Returns a list of attribute indexes found in the input value
ListDatasets(wild_card, feature_type)	Returns the datasets in the current workspace
ListFeatureClasses(wild_card, feature_type)	Returns the feature classes in the current workspace
ListFiles(wild_card)	Returns the files in the current workspace
ListRasters(wild_card, raster_type)	Returns a list of rasters found in the current workspace
ListTables(wild_card, table_type)	Returns a list of tables found in the current workspace
ListWorkspaces(wild_card, workspace_type)	Returns a list of workspaces found in the current workspace
ListVersions(sde_workspace)	Returns a list of versions the connected user has permission to use

List functions

The ListFeatureClasses function will work, but you must set the workspace environment and use a wildcard to get only the box zone files.

6.   Add the workspace code, and the list code using the examples from the tool reference, as shown:

```
# Set up the environment
env.workspace = r"C:\EsriPress\GISTPython\Data\FireDepartment.gdb"
env.overwriteOutput = True

# Create a list of all the box zone feature classes
fcBoxZones = arcpy.ListFeatureClasses("FireBoxMap*")
#fcBoxZones = arcpy.ListFeatureClasses("FireBoxMap_14")
```

The feature class names can now be retrieved from the list object by using an index number, starting at zero (0). The first would be fcBoxZones[0], and the 44th would be fcBoxZones[43] (remember to start the index numbers at 0). Notice that in the graphic an extra line was added to explicitly select number 14. You can use this for testing the script so that it does not have to iterate through all 44 files.

Now use a for statement to retrieve the files one by one. All the code written in the for statement will run for each file in the list object. When the last file is processed, the for statement will release the script to continue to the end.

7. **Write the for statement to go through all the files, and then check your code against the code shown:**

```
# Start a for statement to iterate through the files
for fc in fcBoxZones:
    # Get the first file - it's stored in fc
    print fc
```

8. **Research tools that can create the fields to hold the results. Reference the tool Help, and write the code to add two integer fields named SFCount and MFCount, as shown:**

```
# Add two fields to hold the results
arcpy.AddField_management (fc, "SFCount", "LONG")
arcpy.AddField_management (fc, "MFCount", "LONG")
```

The scenario says that some of the files already have these fields added. The AddField command will automatically detect this and not duplicate a field if it already exists.

Next, get the counts using a select statement. In previous scripts, you copied the feature class to a feature layer and used a query to get only the features you needed. It is interesting to note that multiple feature layers can be created for the same feature class and used in different ways. Consider this option when preparing the select statements.

9. **Add the code to make a feature layer, and perform the selections for single-family homes, as shown:**

```
# Select the single-family housing units (centroid within polygon)
arcpy.MakeFeatureLayer_management("BldgFootprints","Buildings_lyr","\"UseCode\" = 1")
print "Made feature layer"

arcpy.SelectLayerByLocation_management ("Buildings_lyr", "HAVE_THEIR_CENTER_IN",fc)
print "Made selection"

# Count the single-family dwellings
bldgCount = int(arcpy.GetCount_management("Buildings_lyr").getOutput(0))
print str(bldgCount)
```

You have used all three of these statements before with slight differences. Be careful about the syntax for the query in the MakeFeatureLayer command. Also, the selection command uses the wording "HAVE_THEIR_CENTER_IN" for the overlap type parameter so that no building will appear in more than one box. The overlap type keywords must be typed exactly as shown in the tool reference. If you were using the overlap type "INTERSECT," a building that crossed the boundary would be selected twice. Note that some print statements were added so that you can follow the progress of the script as it runs.

Next, update the fields in the feature class to contain the count of single-family homes.

10. Research the tool necessary to put the value bldgCount into the field SFCount, as shown:

```
# Update the field
# CalculateField_management (in_table, field, expression, {expression_type}, {code_block})
arcpy.CalculateField_management(fc,"SFCount",bldgCount)
print "Updated the SFCount field"
```

Your turn

The code for selecting the single-family housing is working. Add the code to do the selection, and update the MFCount field for the multifamily buildings, as shown in the graphic. Remember that the use code for the building footprints is 2 for multi-family.

```
# Select the multi-family housing units (centroid within polygon)
arcpy.MakeFeatureLayer_management("BldgFootprints","Buildings_lyr","\"UseCode\" = 2")
print "Made feature layer"

arcpy.SelectLayerByLocation_management ("Buildings_lyr", "HAVE_THEIR_CENTER_IN",fc)
print "Made selection"

# Count the multi-family buildings
bldgCount = int(arcpy.GetCount_management("Buildings_lyr").getOutput(0))
print str(bldgCount)

# Update the field
arcpy.CalculateField_management(fc,"MFCount",bldgCount)
print "Updated the MFCount field"

# Go to the next feature class, and do the selections again
```

The script is almost complete. The counts will be made for both building types, and the fields will be updated.

11. Save the script. If you were using a single feature class for testing, change the wildcard expression in the list statement to select all the FireBox feature classes. Run the script.

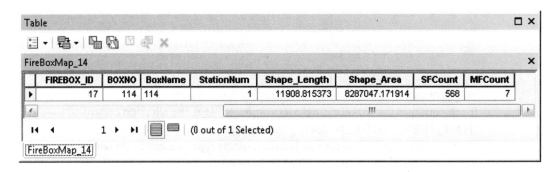

	FIREBOX_ID	BOXNO	BoxName	StationNum	Shape_Length	Shape_Area	SFCount	MFCount
▶	17	114	114	1	11908.815373	8287047.171914	568	7

Here's the pseudo code for this script:

```
# Create a list of all the box zone feature classes
# Start a for statement to iterate through the files
# Get the first file
# Add two fields to hold the results
# Select the single-family housing units (centroid within polygon)
# Count the single-family dwellings
# Update the field
# Select the multi-family housing units (centroid within polygon)
# Count the multi-family buildings
# Update the field
# Go to the next feature class, and do the selections again
```

Exercise 2-5

When you take the results back to the Fire Department, you find the secretary hard at work with a ruler and another printed map. She is trying to measure the linear lane-miles for roads in each box—that is, the length of the street in miles multiplied by the number of lanes. Once again, you tell her that a script can be written to solve this problem and add the results to each of the individual box zone files.

Open the map document Exercise 2-5. The street data and all the box files are shown. Intersect each box file with the streets, and use the fields Shape_Length and Number of lanes to calculate the results into a new field named **LaneMiles**.

Tutorial 2-5 review

As seen in the table of list functions shown in step 5 of this tutorial, there are many list functions in ArcPy. The results of these functions are list objects, which basically contain all the values that are returned. Much like the cursors, a for statement is used to access the values in the list objects. This statement goes through each item in the list one by one and performs the processing tasks you specify. Lists cannot get into an endless loop because they have a finite number of values.

The items in a list object are indexed—for example, text strings in a list. You could pull out an individual item from the list by its index number, if you knew it. Otherwise, you must iterate through the list and find the individual feature class of interest. The list commands also include an optional wildcard value that will let you limit what goes into the list. This may be used to limit the list values to such things as only polygon feature classes or only tables that start with certain letters.

Study questions

1. Select one of the other list types shown in the chart of list functions in the chapter's special introduction, and provide an example of using the list type with a wildcard value.

2. The new fields and final counts were added to the source feature class, but all the selections were done to a feature layer. Why is this desirable?

3. Write an example of code that will return only a single layer from a ListFiles command.

Tutorial 2-6 Building script tools

Scripts can be placed in tools located in a custom toolbox for easy access and easy sharing.

Learning objectives

- Building script tools
- Getting and controlling user input
- Using ArcPy messages
- Creating script tool documentation

Preparation

Research the following topics in ArcGIS for Desktop Help:

- "What is a script tool?"
- "Accessing parameters in a script tool"
- "Understanding messages in script tools"
- "Value list filter"

Introduction

The scripts in the previous tutorials of this chapter all deal with predefined inputs and are designed to run independently of any user input. Although this design functions well in many situations, there are times that you will want to run scripts inside ArcMap and accept user input. To perform these tasks, your scripts must become script tools.

Creating a script tool is straightforward, and several interesting options allow you to make sure things run smoothly. The basic process is to add the script to a toolbox and define the input and output parameters. As with many items in Python, the inputs are indexed in the script, and these index numbers determine the order in which they will appear in the user dialog box. Although there is a Python system command for getting user input, a special function in ArcPy allows you to specify the input type when you create your script tool. This function is named arcpy.GetParamterAsText(), and you increase the index number for each input parameter. Options include defining the type of entry that can be made and pulling a list of values from a domain or attribute table.

Once your script tool is created from a script, it will perform just like the regular ArcGIS tools. You can place it on toolbars, access it from the Catalog window, call it in other scripts, and even index it in the Search window to make it easier to find.

Scenario

You completed a script to perform a single- and multifamily building count for all the box zones in Oleander, and the Fire Department will do periodic counts of other building types in a single box on an as-needed basis. With some simple modifications, you can have the script get input from the user to decide what box zone to work on and which building type to search for. The resulting building count can be stored in a new field with a predetermined name and then displayed in the geoprocessing Results window. Because the building count will differ as the city changes, it will be helpful to run this script to get fresh counts each time.

Data

The data is the box map layer and the individual box zone files from the City of Oleander geodatabase. Also in the map document are the building footprints. The field UseCode has a numeric code for each use type. The following list contains the use type followed by the field name you should use to store the building count for each type:

1 = Single Family (SFCount)
2 = Multi-Family (MFCount)
3 = Commercial (ComCount)
4 = Industrial (IndCount)
5 = City Property (CityCount)
6 = Storage Sheds (ShedCount)
7 = Schools (SchCount)
8 = Church (ChurCount)

SCRIPTING TECHNIQUES

The new technique shown here is how to create a script tool from a script. Two important steps will help lead to predictable results. The first step is to carefully track any input or output variables. Make notations in the code to identify the order in which the input values are accepted from the user. The script tool must consume these values in the same order. The index numbers are used to track this order, starting with zero and moving up from there.

Although this script does not use it, there is a special function for returning a value from a script tool. This function is primarily used to feed a return value to a model, but it may also be used to send a value to another script or model. This function is arcpy.SetParameterAsText(), which is tracked with an index number and is set as output in the script tool.

The second step is to ensure that the input value type is set correctly. Setting these value types will cause the script to use the standard ArcMap input dialog boxes, giving your custom tool the look and feel of a system tool. But remember that whatever type you set, the dialog box will expect you to select that type of item. For instance, if you set the input to Feature Class, the dialog box will have you select an existing feature class. But if the tool is set to create a new feature class, you will not be able to give it a valid name because you will be prompted to select an existing feature class. In this case, you would set the input to a string, and then use the string to create the feature class. However, if you are setting a default workspace, and you set the input to workspace, only valid, existing workspaces can be selected, which would be correct. Use the rule that whatever data type you set, an existing item of that data type will be selected.

Another interesting technique is the use of an input filter. You can set a filter for certain data input types. This filter can be as simple as a predefined list of values, or it may include complex code, which you will learn in tutorial 2-7. The list will limit the choices of the user—these types of data integrity rules are always a good idea. In addition to the list, you will add documentation to the script so that the user can better understand how to use the tool—another good practice. Make it a goal with all your scripts to document the code thoroughly for future programmers, set as many data integrity rules as possible to make it easier on the user, and add as much detailed documentation as possible to avoid user confusion.

Build script tools

1. Open the map document Tutorial 2-6. The box zones and the building footprints are as shown:

2. Examine the data, and determine your course of action. Write your pseudo code outlining the necessary steps. Completed pseudo code is shown at the end of this tutorial.

3. Start a new script in your IDE named Tutorial2-6.py. Although this script is being written to your IDE, it will later become an ArcGIS script tool. Set up the template lines in your script to store the description, and load the ArcPy module. Set the workspace environment to C:\EsriPress\GISTPython\Data\City of Oleander.gdb.

```
#-----------------------------------------------------------------
# Name:        Tutorial 2-6
#              Dwelling count with user input
# Purpose:     Prompt the user for the building type and the box number.
#              Count the buildings in the selected box zone and store the results
#              in the attribute table.
#
# Author:      David W. Allen, GISP
#
# Created:     10/25/2013
# Copyright:   (c) David 2013
#
#-----------------------------------------------------------------

try:
    # Import the modules
    import arcpy
    from arcpy import env

    # Set up the environment
    env.workspace = r"C:\EsriPress\GISTPython\Data\City of Oleander.gdb"

    # ENTER CODE

    # Determine results

except arcpy.ExecuteError:
    print arcpy.GetMessages(2)
except:
    print "Process did not complete."
```

You must prompt the user for the box number and the building type. Use the arcpy.GetParameterAsText() function with index numbers 0 and 1, but later when this becomes a script tool, you will see how to make input statements a list type input.

4. Write the code to get the box zone number as a variable named boxNumber and the building type code as a variable named buildingType with index numbers 0 and 1, respectively, as shown:

```
# Get input from the user
#     The first will be the box number to act upon - index 0
#     The second will be the building type to count - index 1
boxNumber = arcpy.GetParameterAsText(0)
buildingType = arcpy.GetParameterAsText(1)
```

Remember these index numbers because they must be listed in the same order in the script tool. The first input will be the box number in which the user wants to perform a count, and the second input will be the building type to use for the count.

5. Use the MakeFeatureLayer tool to get the correct box zone file and the proper set of building footprints, as shown:

```
# Make feature layers from the user input
boxLayer = arcpy.MakeFeatureLayer_management(\
r"C:\EsriPress\GISTPython\Data\City of Oleander.gdb\FireBoxMaps\FireBoxMap_" \
+ str(boxNumber))
buildLayer = arcpy.MakeFeatureLayer_management(\
r"C:\EsriPress\GISTPython\Data\City of Oleander.gdb\Planimetrics\BldgFootprints", \
"\"UseCode\" = '" + buildingType + "'")
```

Note that because you are making a variable equal to the MakeFeatureLayer tool, you do not need to add an output layer name to the tool's syntax.

Do the selection and count just as you did in tutorial 2-5.

6. Write the code to select the buildings based on the specified box zone, and count the number of selected features, as shown:

```
# Use the specified file of box zone to select specified type of building
arcpy.SelectLayerByLocation_management(buildLayer, "HAVE_THEIR_CENTER_IN", boxLayer)

# Count the selected features
bldgCount = int(arcpy.GetCount_management(buildLayer).getOutput(0))
```

This code will get the building count. Write this code to the geoprocessing Results window, and note that there is a special ArcPy function to do this—in fact, there are three. The differences are basically in the color of the text when the message is printed. The commands are as follows:

- arcpy.AddMessage, which adds text to the Results window in black letters
- arcpy.AddWarning, which adds text to the Results window in green letters
- arcpy.AddError, which adds text to the Results window in red letters

Note that none of these commands will interrupt the script, but each will print a message in its associated color. Next, use one of these commands to add a statement about the feature count to the Results window.

7. Add the message command to put the feature count in the Results window, as shown:

```
# Display the results in the geoprocessing Results window
arcpy.AddMessage("The count of buildings is" + str(bldgCount) + ".")
```

Next, create a field to store the results, and calculate it equal to the count. The field name will depend on the building type the user specified. Remember that the ArcPy AddField function will work, even if the field already exists.

8. Write code that uses the buildingType variable to determine the proper field name, and add it to the selected feature class. Then calculate the count value to that field, as shown:

```
# Create a field to store the results

# 1 = Single Family (SFCount)
# 2 = Multi-Family (MFCount)
# 3 = Commercial (ComCount)
# 4 = Industrial (IndCount)
# 5 = City Property (CityCount)
# 6 = Storage Sheds (ShedCount)
# 7 = Schools (SchCount)
# 8 = Church (ChurCount)

if buildingType == 1:
    newField = "SFCount"
elif buildingType == 2:
    newField = "MFCount"
elif buildingType == 3:
    newField = "ComCount"
elif buildingType == 4:
    newField = "IndCount"
elif buildingType == 5:
    newField = "CityCount"
elif buildingType == 6:
    newField = "ShedCount"
elif buildingType == 7:
    newField = "SchCount"
else:
    newField = "ChurCount"

arcpy.AddField_management(boxLayer,newField,"LONG")

# Store the results in the field
arcpy.CalculateField_management(boxLayer,newField,bldgCount)
```

This graphic shows a lot of code, but it is basically three steps: determine the field name to use, add that field, and calculate the value.

9. **Now that your code entry is complete, save your Python script to your MyExercises folder where you installed the student data, and close your IDE.**

Here's the pseudo code for the script:

```
# Import the modules, and set the environments
# Get input from the user
#    The first will be the box number to act upon
#    The second will be the building type to count
# Make feature layers from the user input
# Use the specified file of box zone to select specified type of building
# Count the selected features
# Display the results in the geoprocessing Results window
# Create a field to store the results

# 1 = Single Family (SFCount)
# 2 = Multi-Family (MFCount)
# 3 = Commercial (ComCount)
# 4 = Industrial (IndCount)
# 5 = City Property (CityCount)
# 6 = Storage Sheds (ShedCount)
# 7 = Schools (SchCount)
# 8 = Church (ChurCount)

# Store the results in the field
```

Script tools must be stored in an existing toolbox, so create a new one in your MyExercises folder.

10. **In ArcMap, open the Catalog window, and navigate to your MyExercises folder where you installed the student data. Right-click the folder, and click New > Toolbox, as shown in the graphic. Name the toolbox** Custom Python Tools.

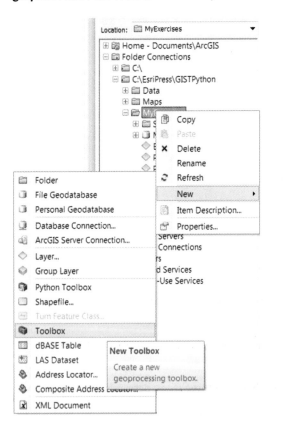

11. Right-click the toolbox, and click Properties. For an alias, type custompython, and add a description, as shown in the graphic. Click OK.

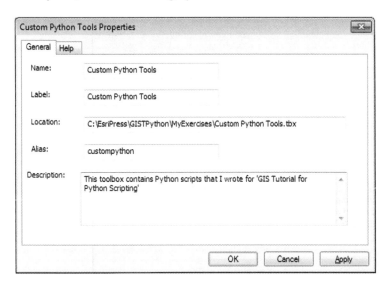

Adding aliases to your toolboxes is important. When you access tools from a toolbox in a Python script, use the tool name and an underscore and then the toolbox alias. Have you noticed that the ArcPy tools you have been using all have a toolbox alias after the tool name? Using the alias helps distinguish between two tools that might have the same name. You will not use the alias here, but it is good practice to assign one for later use if you index your custom toolboxes.

Start the process by adding your script to the toolbox. There are several steps, so be careful not to miss one.

12. Right-click the Custom Python Tools toolbox and click Add > Script, as shown:

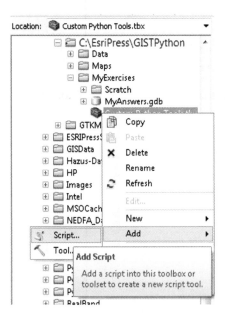

This opens the script tool dialog box, and you can set the parameters.

13. Name the tool CountBuildings, label it Count Buildings, and give the tool an appropriate description. Also, select the "Store relative path names" check box, as shown in the graphic. Click Next.

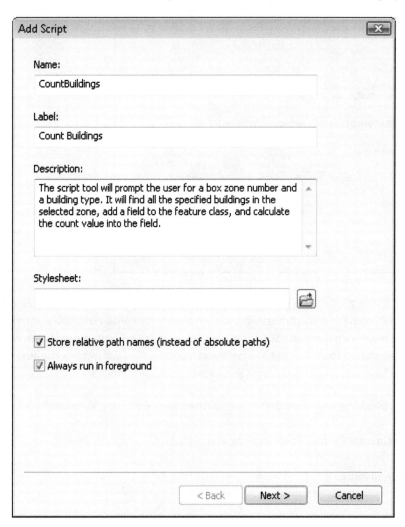

14. Click the Navigate button, and select your script from your MyExercises folder, as shown in the graphic. Click Next.

Next, define the input parameters for the two values you need from the user. Remember that the index numbers you assigned will get the box zone number first and the building type second.

15. As shown, click the first cell under Display Name and type Box Zone Number, which will act as a text prompt to the user.

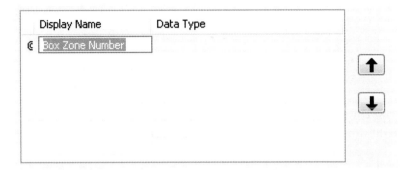

Setting the data type is next. When you click the Data Type arrow, a long list of data types appears. It is important to realize that if you choose a particular type of file, such as a feature class, shapefile, or dBASE file, the input must be a file of this type and must already exist. The advantage of setting the data type is that the selection dialog box will only take files of the specified type. For instance, if you set the data type to Workspace and navigate to C:\EsriPress\GISTPython\Data, you will see only geodatabases listed and none of the text files or shapefiles that might be there. If you are just getting text that will be used to create one of these types of files, your input should be a string. For this input, you can select String to allow alphanumeric entries.

16. Click the Data Type box and select String, as shown:

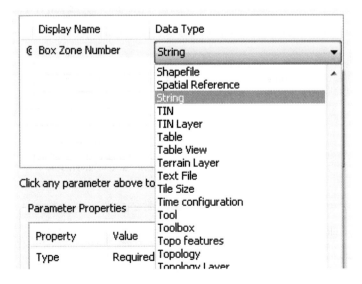

17. On the next line, type the display name as Building Type, and set it to String, as shown in the graphic. Note that in the Parameter Properties box both of these items are set as Required and Input.

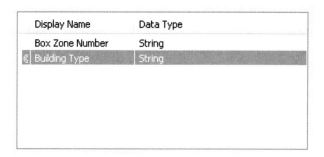

18. With Building Type still highlighted, move to the Parameter Properties panel, and click the cell next to Filter. In the drop-down list, select Value List, as shown:

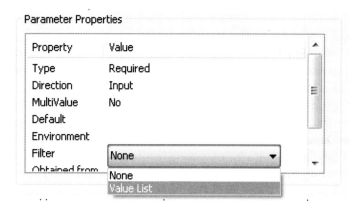

This dialog box will let you build a value list to make data entry easier for the user. This value list can also prevent users from entering an invalid value. The table lists the filter types you can apply with this setting, as shown:

Filter type	Values
Value List	A list of string or numeric values. Used with String, Long, Double, and Boolean parameter data types.
Range	A minimum and maximum value. Used with Long and Double data types.
Feature Class	A list of allowable feature class types: "Point", "Multipoint", "Polyline", "Polygon", "MultiPatch", "Sphere", "Annotation", and "Dimension". More than one value can be supplied to the filter.
File	A list of file suffixes, for example, "txt; e00; ditamap".
Field	A list of allowable field types: "Short", "Long", "Single", "Double", "Text", "Date", "OID", "Geometry", "Blob", "Raster", "GUID", "GlobalID", and "XML". More than one value can be supplied to the filter.
Workspace	A list of allowable workspace types: "File System", "Local Database", or "Remote Database". More than one value can be supplied.

19. In the Value List, enter the numbers 1 through 8, as shown in the graphic. Click OK, and then click Finish to complete the script tool creation process.

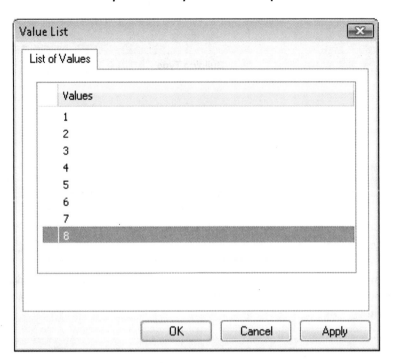

20. Run the script tool by double-clicking it in the toolbox. Enter a box zone number of 100, and select a building type of 5, as shown in the graphic. Click OK to run the tool, and make sure to clear the check box "Close this dialog when completed successfully." If the dialog box closes too quickly, or if the script is running in the background, you can open the geoprocessing Results window to see the building count.

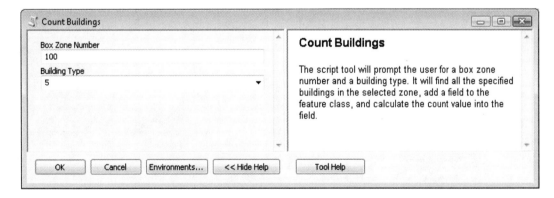

If you completed the steps correctly, the tool will run successfully. If the tool fails, right-click the tool in the toolbox and click Edit. This opens an editing tool that lets you make changes to the script. The default editing tool is Notepad. Refer to appendix A for how to set the default editing program for Python scripts used in ArcGIS.

Did you notice that the description you typed earlier during the tool creation process became the tool Help? You may have also noticed that when you clicked the entry boxes to type values, you did not get a Help message, as shown:

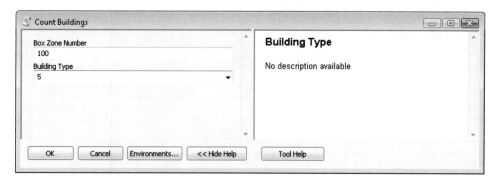

There is a way to add more relevant help to the tool, enhance its appearance, and give it the same look and feel as a system tool.

21. **Right-click your new tool and click Item Description. In the resulting window, click Edit. Scroll through the window, and note the variety of items that can be entered (and that tags are required), as shown:**

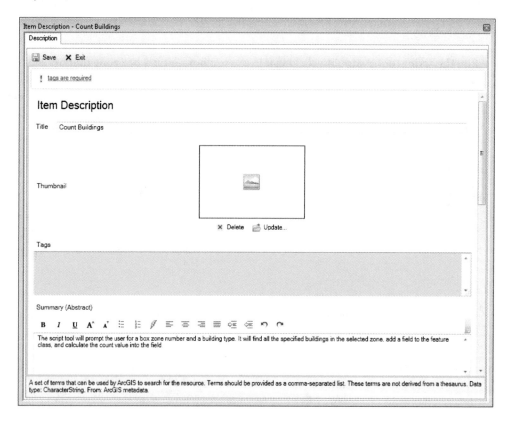

As you click each of the entry lines, a small Help text is displayed at the bottom of the window to guide you toward inputting the correct information.

22. Scroll down and click the chevron next to Box_Zone_Number. Here, you can type more descriptive instructions for the user. Notice that there are many formatting options, including bold, italic, and bullet lists. Fill in detailed instructions, as shown:

23. Add a better description for the parameter Building_Type, as shown:

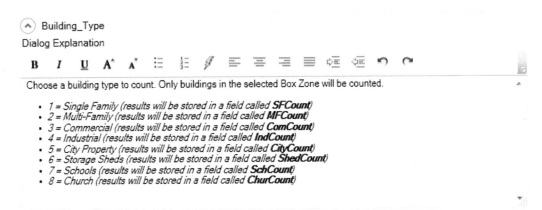

24. Explore the other description items, and enter tags that will identify the purpose of this script. When you are done, click Save and close the Item Description dialog box.

25. Run the script again, and click in each of the entry boxes to see the resulting Help messages, as shown:

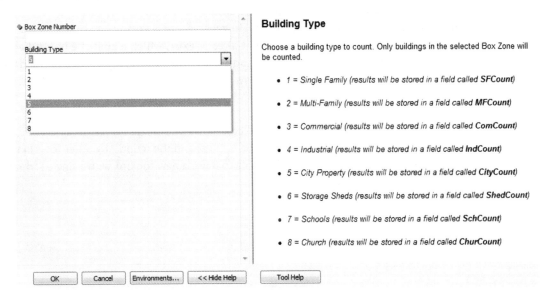

The additional Help creates a better user experience.

In this script, you are having the user enter a number to represent the building type. As a bonus, can you alter the script to allow the user to select a text building type description in the input box, and then can you have the script figure out what value to use and type that text description in the results message?

Exercise 2-6

Use the script you wrote in exercise 2-5 to create a script tool. Make changes to allow for user input, and develop a rich Help environment to make data entry easy to understand.

Tutorial 2-6 review

Creating script tools introduces another aspect of ArcGIS customization. With the other scripts you have created, the user ran them from a folder and so could not appreciate how they interacted with the map. The user also did not have an opportunity to change the input feature classes unless they could edit them in an IDE, which is not exactly the most user-friendly experience. Also, no Help file existed to instruct the user on the correct use of the tool.

With a script tool, you are creating an item that can be run from the Catalog window, or even moved onto a toolbar. Script tools provide the rich navigation environment in ArcGIS that allows you to do such things as select layers from the table of contents and filter the navigation to feature classes. You also have the option to write custom Help files for the tool, which will help inform the user how to use the tool.

When the tool was created, you had to specify the data type for each input. A long list of available data types to choose from exists, and you should become familiar with this list. You may see many sample scripts using the data type of Any Value, which is fine for naming items that do not currently exist. However, there is an advantage to using a data type where the user must select items that already exist, which is that the input dialog box will be the same one that ArcGIS system tools use. For instance, if you specify the data type as Table, the user input dialog box will show only table format files as the user browses through folders.

Script tools must be created in a custom toolbox, either one created earlier or a new one. To share script tools with others, you must provide the toolbox and the script. Any other tools in the toolbox will also be shared, so it is sometimes advisable to make a new toolbox with a single script tool when you are sharing scripts.

Study questions

1. Look over the list of data types for script tool inputs in the script tool parameters Data Type drop-down list, and give three examples of where specific data types should be specified. Describe what the input dialog box should look like for each data type example you have chosen.

2. The filter option was used to restrict the input for the building types. Give examples of other uses of the filter.

3. Explain why the context-sensitive Help is important.

Tutorial 2-7 Using cursors

Cursors are a programming technique used to step through feature classes and tables, item by item. Each item the cursor finds can be altered independently or used in a geoprocessing task.

Learning objectives

- Working with table properties
- Using cursors with tables
- Using input validation in script tools

Preparation

Research the following topics in ArcGIS for Desktop Help:

- "Make Table View (Data Management)"
- "Understanding validation in script tools"

Introduction

The lists and cursors that you have used up until now have worked mostly with geodatabases and feature classes, but these techniques can also be used on tables. You may want to get a list of tables or use a cursor to go through the rows of a table, one by one.

As with the cursors you have used on features, cursors applied to rows in a table also allow individual processing. You can access the fields and their values and then use decision-making and condition statements to perform a variety of tasks.

Scenario

On your trip to the Fire Department to show off the building count application, you discover another complex process that the department is trying to perform manually (you really need to stop going over there). In any given month, the Oleander Fire Department responds to several hundred calls around the region. Since Oleander has mutual-aid agreements with 16 neighboring cities, the call location may be in any one of these locations. The department needs to split the fire run data that represents each call for service into multiple files with one table for Oleander calls and another table for each of the other cities in which they responded.

The department would also like to geocode the calls, but in the current format, the addresses are parsed into separate fields, so the addresses will need to be combined. You could reformat the address data in the main file, but that would mess up the access to the historical data in the automated dispatch software (the field schema cannot change). This change can only appear in the output files, and not in the source data.

You also cannot use a simple query to split this data because you do not know what cities to use in a selection. The list of cities may be different each month. However, you should be able to make a list of all the city names and use that list to do your table selections.

The solution is to create a temporary copy of the table in memory, make any field changes you need, and calculate the new address field. Use the text slicing technique and a file creation tool to create a new geodatabase that includes the name of the month and the year. Then use a cursor to run through the data to develop a list of unique city names. Use that list to do selections on the database, and create new tables for each unique city name that is found.

You can run this script from ArcCatalog since there are no graphics involved. The tools necessary are listed here so that you can do some research before writing your pseudo code:

MakeTableView

SelectLayerByAttribute

CreateGeodatabase

CopyRows

CalculateField

FieldInfo()

.replace()

The pseudo code will be complex, so think through the process as if you were doing only one record and document it. Then check to see what other events might occur to change the process, and accommodate for these events in your pseudo code. None of the tasks by themselves are that difficult, but combining them will take planning and precision.

Data

The only data used for this tutorial is the run data for each time frame. A couple of examples are given, but once the script is written, it could be used for years to come to perform the same function on any new dataset created from the dispatch software.

SCRIPTING TECHNIQUES

This tutorial combines many of the techniques you have learned in other tutorials. This tutorial uses cursors, table views, for statements, and several different ArcGIS tools. A new technique to try is the use of a with block to combine the cursor and the for statements into a single subset of the code. The format is as follows:

```
with {cursor setup} as {cursor name}:
    for {name to track current row} in {cursor name}:
        # Add your code here to process cursor items
```

There are two major advantages of using a with block. The first advantage is that the cursor and the for statements are defined in two lines of code. The second advantage is that references to any files or map documents in the with block are automatically removed at the end of the code block, even if the code fails, which prevents a crashed script from locking those datasets or map documents.

Another new technique is the use of input validation code. By adding this extra code in the input properties, the programmer can greatly control what the user is allowed to use for input into the script. This technique is another type of data integrity rule that can range from checking for the correct file name to examining the field structure before allowing the selection. Its function is dependent on the validation code you write.

Use cursors

1. **Start ArcCatalog. Research the tools you may need, and write your pseudo code. A set of pseudo code is shown at the end of this tutorial. If your pseudo code differs substantially, try using it for writing the code, and reference the tools as presented in these steps.**

2.   **Start your IDE, and set up the template lines for a new script named** ProcessFireData**. The workspace environment for outputting new files is** C:\EsriPress\GISTPython\MyExercises**, as shown:**

```
try:
    # Import the modules
    import arcpy
    from arcpy import env

    # Set up the environment
    env.workspace = r"C:\EsriPress\GISTPython\MyExercises\\"
    env.overwriteOutput = True
```

The process starts by asking the user which file they want to work on. In the code, you can use the generic GetParameterAsText() function, and when you create the script tool, set up the data type.

3.   **Add the line of code to accept user input into a variable named** inTable**, as shown:**

```
    # Prompt user for the input table
    inTable = arcpy.GetParameterAsText(0)
    # When this is set up as a cursor tool, set the input to tables only
```

Next, make a copy of the input table in memory with the MakeTableView command. This copy allows you to make selections and queries without affecting the source data, unless you actively save the table view object. You do not need to show all the fields in the output, but you must add a field that can be used to store the concatenated address.

ArcGIS for Desktop Help has a great example of how to do this. Note that you cannot add a field to a table view, but the sample code in the tool reference for MakeTableView shows a way to change the name of a field that is not being used.

First, get a list of all the fields in the input table, and then create an object that will store the information about the fields. This object will be modified and used in the MakeTableView command later on.

4.   **Get a list of all the fields in inTable, and store the list in a list object named** fields**. Then use the FieldInfo() function to create a field object, as shown:**

```
    # Get the fields from the input
    fields = arcpy.ListFields(inTable)

    # Create a fieldinfo object
    fieldinfo = arcpy.FieldInfo()
```

Before you create the table view, alter the field information for the output. This alteration will make certain fields visible in the output and will change the name of addr_2 to GeoAddress, which will later accept the concatenated address string. At the end of the elif statements is an else statement that hides all unidentified fields in the output. The tool reference for MakeTableView has a good example of this type of statement.

5. Add condition statements to make the list of desired fields visible in the table view. Also, change the name of addr_2 to GeoAddress, as shown:

```
# Define a fieldinfo object to bring only certain fields into the view
#  inci_no, alm_date, alm_time, arv_date, arv_time, inci_type
#  descript, station, shift, city
#  number, st_prefix, street, st_type, st_suffix
#  (you can't add new fields to a table view, so reuse a discarded one)
#  Change the name of addr_2 to GeoAddress in the output table
# Code was copied and modified from the Help screen

# Iterate through the fields, and set them to fieldinfo
for field in fields:
    if field.name == "inci_no":
        fieldinfo.addField(field.name, field.name , "VISIBLE", "")
    elif field.name == "alm_date":
        fieldinfo.addField(field.name, field.name , "VISIBLE", "")
    elif field.name == "alm_time":
        fieldinfo.addField(field.name, field.name , "VISIBLE", "")
    elif field.name == "arv_date":
        fieldinfo.addField(field.name, field.name , "VISIBLE", "")
    elif field.name == "arv_time":
        fieldinfo.addField(field.name, field.name , "VISIBLE", "")
    elif field.name == "inci_type":
        fieldinfo.addField(field.name, field.name , "VISIBLE", "")
    elif field.name == "descript":
        fieldinfo.addField(field.name, field.name , "VISIBLE", "")
    elif field.name == "station":
        fieldinfo.addField(field.name, field.name , "VISIBLE", "")
    elif field.name == "shift":
        fieldinfo.addField(field.name, field.name , "VISIBLE", "")
    elif field.name == "city":
        fieldinfo.addField(field.name, field.name , "VISIBLE", "")
    elif field.name == "number":
        fieldinfo.addField(field.name, field.name , "VISIBLE", "")
    elif field.name == "st_prefix":
        fieldinfo.addField(field.name, field.name , "VISIBLE", "")
    elif field.name == "street":
        fieldinfo.addField(field.name, field.name , "VISIBLE", "")
    elif field.name == "st_type":
        fieldinfo.addField(field.name, field.name , "VISIBLE", "")
    elif field.name == "st_suffix":
        fieldinfo.addField(field.name, field.name , "VISIBLE", "")
    elif field.name == "addr_2":
        fieldinfo.addField(field.name, "GeoAddress" , "VISIBLE", "")
    else:
        fieldinfo.addField(field.name, field.name, "HIDDEN", "")
```

When you create the table view, it will contain only the fields you want.

6.   Create a new table view using the user's input table and the field object you defined, and name it
fire_view, **as shown:**

```
# Create a table view of the input table
# The created fire_view table view will have fields as set in fieldinfo object
arcpy.MakeTableView_management(inTable, "fire_view", "", "", fieldinfo)
```

7.   **Write the code to reformat the address fields into a single field named** GeoAddress, **as shown.**
This code is similar to the code you wrote in chapter one.

```
# Do the address formatting into GeoAddress for the whole table
# Concatenate number + st_prefix + street + st_type + st_suffix and remove spaces
arcpy.CalculateField_management("fire_view","GeoAddress","str(!number!) + ' ' +\
!st_prefix!.strip() + ' ' + !street!.strip() + ' ' + !st_type!.strip() + ' ' +\
!st_suffix!.strip()", "PYTHON")
```

Note that this reformatting only makes changes to the table view and does not change the source
data unless that change is specifically intended.

The output of the process must be stored in a new table in a new geodatabase. To get the year of the
data's collection, slice the last four digits from the input file name. Then add the year as a suffix to a
new geodatabase name, and store the geodatabase in your MyExercises folder.

8.   **Use string slicing to get the year for the input data. Then create a new geodatabase with the year**
as the suffix for the file, as shown:

```
# Create new geodatabase to store results for year
# ("Fire Files for " + last 4 digits of file name)
gdbName = "Fire_Files_For_" + inTable[-8:]
arcpy.CreateFileGDB_management("C:\\EsriPress\\GISTPython\\MyExercises",gdbName)
```

The eventual goal is to create a separate database for each of the different city names
containing only the records for each city, but you do not know what those city names will be.
In some months, there may be only four or five cities with responses, but in a busy month,
the number can grow to as many as 14. You can find all the values by having the cursor iterate
through the records to look for unique names, and then store the values in a list variable.
Remember that a list variable can store multiple values, which are retrieved using an index
number.

You can initialize an empty list variable by making it equal to a set of empty brackets ([]). As you
find the unique values, they can be determined to be in the list and, if necessary, appended to
the list. Check your Python reference for the syntax of these functions. A with statement can set
up the cursor and the framework for the iteration. Then a for statement can run through all the
records, and an if statement can determine whether the found value of cityName is already in
the list.

9. Add the code to set up the with block, which will scroll through the rows looking for city names, and create a list variable to store the unique results. If you like, add an ArcPy message to show that the list was created successfully, as shown:

```
# Use cursor to find each unique city name and add it to a list.
# City names included may differ from file to file.
# Set up a list to hold unique city names.
cityList = []

# Start cursor iteration
for row in fireCursor:
    cityName = row[23]
    if cityName not in cityList:
        cityList.append(cityName)
# Result is a list object with all the unique values of the CITY field
del row
arcpy.AddWarning("Made the list of city names.")
```

You are ready to create the new tables. By going through the new list of city names, you can get both the value to use in a search of the table view and the name of the new output table.

10. Use a for statement to go through the city names list, select all the records from that city, and write a new table to the geodatabase you created in step 8, as shown:

```
# Use the names in the list object to select records
for name in cityList:
    cityQuery = '"city" = \'' + name + '\''

    arcpy.SelectLayerByAttribute_management ("fire_view", "NEW_SELECTION",cityQuery)

    newTable = "C:\\EsriPress\\GISTPython\\MyExercises\\" + gdbName + ".gdb\\" + \
    name.replace(" ","_")

    arcpy.CopyRows_management ("fire_view", newTable)

    itemCount =int(arcpy.GetCount_management("fire_view").getOutput(0))

    arcpy.AddWarning("A table called " + newTable + " was created with " + \
    str(itemCount) + " rows.")
```

Note that when the new table is created, the spaces in the city names are replaced with underscores because ArcMap does not accept a space or a special character in the name of a table. Notice also that code was added to count the number of features and to report the names and record counts of the new tables back to the Results window as the tables are created. Although this action is not required, it is helpful to let the user know what the script is doing.

11. Save the script, and close your IDE.

The script takes only one input—the fire department calls for service—so your script tool will need only one input, which you can restrict to tables.

12. Start ArcCatalog. In the Catalog window, create a new toolbox in your MyExercises folder named Tutorial 2-7.tbx. **Right-click the toolbox and click Add > Script. As shown, name the tool** SplitFireCalls**, and add an appropriate label and description. Select the "Store relative path names" check box and click Next.**

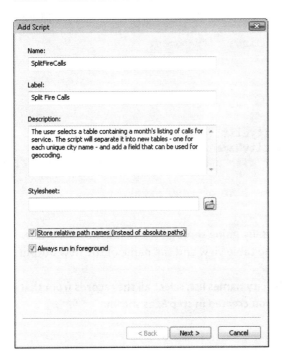

13. Navigate to the folder where you stored the script and select it. Click Next.

14. Set the display name to Calls for Service **and the data type to Table, as shown. Click Finish.**

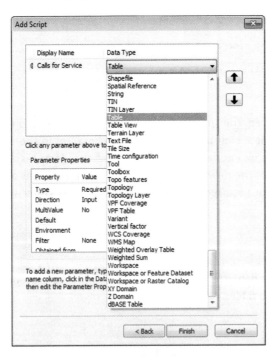

15. Run the script tool, and test it with the January run data in C:\EsriPress\GISTPython\Data\ FireDepartment.gdb, as shown. Debug the script tool, if necessary.

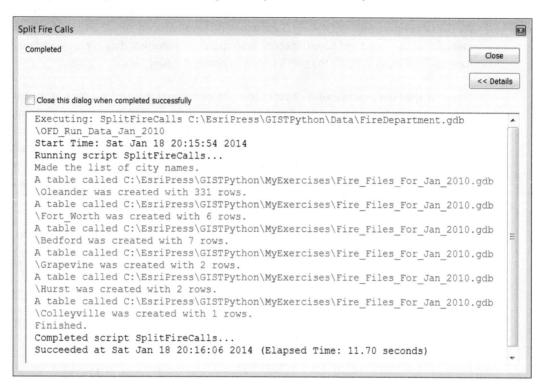

Here's the pseudo code for this script:

```
# Import the modules.
# Set up the environment.
# Prompt user for the input table.
# When this is set up as a cursor tool, set the input to tables only
# Get the fields from the input.
# Create a fieldinfo object.
# Define a fieldinfo object to bring only certain fields into the view.
#   inci_no, alm_date, alm_time, arv_date, arv_time, inci_type
#   descript, station, shift, city
#   number, st_prefix, street, st_type, st_suffix
#   (you can't add new fields to a table view, so reuse a discarded one)
#   Change the name of addr_2 to GeoAddress in the output table
# Code was copied and modified from the Help screen.
# Iterate through the fields, and set them to fieldinfo.
# Create a table view of the input table.
# The created fire_view table view will have fields as set in fieldinfo object.
# Do the address formatting into GeoAddress for the whole table.
# Concatenate number + st_prefix + street + st_type + st_suffix and remove spaces.
# Create new geodatabase to store results for year ("Fire Files for " + last 4 digits of file name).
# Create a cursor to go through the table view row by row.
# Use cursor to find each unique city name, and add it to a list.
# City names included may differ from file to file.
# Set up a list to hold unique city names.
# Start cursor iteration.
# Result is a list object with all the unique values of the CITY field.
# Use the names in the list object to select records.
# Repeat for all names in the list.
# Use the script to create a script tool.
# Add validation code to the script tool.
```

The tool runs fine and creates the required outputs, but there is still one problem: the user could accidentally enter a table with the wrong formatting, and the script would fail. It would be helpful to test the input file first and make sure it contains the correct type of data in the correct format. All the tables start with OFD_Run_Data, which makes them easy to identify. Rather than put this identifier in the script and have the script stop running if incorrect data is provided, you can instead have the tool input validate the user's selection before the tool runs.

You have seen this practice with system tools: the user enters a value for a parameter, and a red *X* appears next to the entry and prevents the tool from running. This preventative technique is called *tool validation* and is available for use on custom scripts as well. The tool validation code is accessed through the script properties and allows you to check the user's input as values are entered, change the tool parameters based on the input, and change output messages based on certain conditions. For this script, add a check at the tool's initialization to make sure the first 12 characters are OFD_Run_Data before allowing the user to click OK.

16. Right-click the new script tool SplitFireCalls, click Properties, and click the Validation tab.

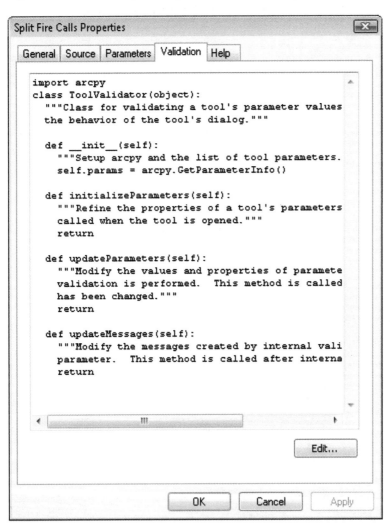

The Validation tab shows the standard Python code for the validation object associated with the tool. All the code here is required; you can add to it, but you cannot delete anything. The object created is named *self* and has an index to each parameter value that the toolbox asks for, starting with zero (0). For example, the first parameter would be self.params[0], and the fifth parameter would be self.params[4]. This object has methods, which can be used to raise an error, and parameters, such as value, data type, and a True/False flag, to identify whether the value has been altered in the input box. Using these methods and parameters, you will be able to control the data entry with more precision than with the normal filter in the script tool entry box.

ArcGIS for Desktop Help has more information on tool validation under "Programming a ToolValidator class." Research this topic, and follow along with the code that follows.

The first step is to get the input value from the user interface. When the script detects that the input value has been changed, you can check its file name to make sure that the first 12 characters use the proper naming convention. If not, an error message appears, blocking the user from continuing.

17. Click Edit to open your IDE with this script. Under updateMessages, add the following code:

```
def updateMessages(self):
    """Modify the messages created by internal validation for each tool
    parameter.  This method is called after internal validation."""
    if self.params[0].altered:
        inputValue = arcpy.Describe(self.params[0].value)
        if not inputValue.file[:12] == "OFD_Run_Data":
            self.params[0].setErrorMessage("Wrong file." +\
            " Select a file containing Oleander Fire Department run data.")
    return
```

Note that the code is first using the .altered property to check whether the user entered anything. If so, it describes the data into a variable named inputValue. Using the arcpy.Describe method slices the input value into its various components, including the path, file name, and extension. Using the file property and slicing the first 12 characters, the code then checks that the file uses the correct naming convention. If it does not, an error message is returned, which adds the red *X* on the input screen and lets the user know that the entry is not valid.

18. Close your IDE, and save the code. In the Script Properties dialog box, click Apply. Your code now appears in the validation code block. Click OK. Run the script tool with both a valid and an invalid entry. Shown is the error message that appears with an invalid entry:

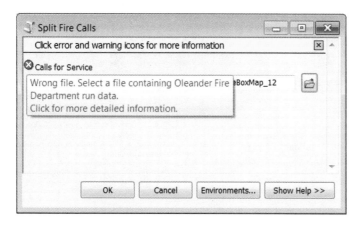

19. Alter the script's description to add useful help in the standard tool Help dialog box.

Using the tool validation in conjunction with the tool validation code is a fantastic way to put custom restrictions on user input and to prevent users from accidentally running a tool incorrectly. Research this topic more in ArcGIS for Desktop Help to fully understand how much control you have.

Exercise 2-7

After the Fire Department separates the data by city, it would like to select only the Oleander data and separate it into the different incident types (inci_type). Again, no set list exists of what incident types the file will have, and the output file name should indicate the month and year of the data.

Create a separate script tool to reprocess the results of the Split Fire Calls tool to split the Oleander calls from the resulting file.

Add controls in the tool validation code to make sure the correct file is chosen before the tool runs.

Tutorial 2-7 review

This application uses a variety of techniques learned from previous tutorials and introduces the idea of input validation, which can be extremely important in script tools that might be used by a large number of users who may possess different levels of GIS knowledge. This concept is known as data integrity—the use of code to keep users from creating errors in your data and to help them make correct choices. The more data integrity techniques you can incorporate in your applications, the easier it will be to maintain the integrity of your dataset.

Study questions

1. Make a list of the techniques used by this application that you learned from other tutorials.
2. Give examples of other uses of input validation code.
3. Which input data types allow for filters and validation code?

Tutorial 2-8 Combining loops

Any of the looping techniques, such as while and for statements, can be combined to provide a flexible method of working with data in Python.

Learning objectives

- Validating data
- Summarizing data
- Managing flow of processing

Preparation

Research the following topics in ArcGIS for Desktop Help:

- "Summary Statistics (Analysis)"
- "Get Count (Data Management)"

Introduction

By completing all the previous tutorials in this book, you have learned many different analysis and programming techniques. These techniques include branching using if statements, for and while loops, cursors, and lists. This tutorial uses many of these techniques in a single script tool and requires investigation into how the analysis processes work, how the user will be using this script tool, and what tools and techniques you will need to incorporate into your script tool. This tutorial also requires detailed pseudo code, which you should be an expert at by now.

Scenario

The Oleander Public Works Department in the Engineering division is gearing up for a new storm water testing program to make sure that the water being discharged into the streams and rivers meets the federal clean-water standards. You already have a dataset of the storm drain system, and thanks to the department's summer intern, who slogged up and down every creek in Oleander with a GPS unit, you also have the locations of all the outfalls. These locations are the points at which a storm drain collection system for a particular watershed empties into the creek, and there are many of these outfalls. The department has set up monitoring stations at a few of the locations to start taking water quality readings, and a system name has been given to each station. To complete the analysis, the department needs to know the characteristics of the drainage system for the watershed connected to the particular outfall at the monitoring station. These descriptions need to include an inventory of the fixtures attached to the watershed system, along with a summary of pipe sizes. From this information, the department can calculate maximum capacity and determine whether the readings at the outfalls are within specifications.

The script tool you will write to automate the process will have the user select one outfall, and then the tool will trace the connected pipes until all are selected. This tool can be used to create the pipe inventory. Then the tool will select and form an inventory of the fixtures attached to these pipes. A data maintenance task can also be completed while you're at it. The fixtures are supposed to have a value for the size pipe they are connected to, but that data was not populated when the storm drain database was created. This problem will be easy for your script to solve as it goes through the features.

Data

The data provided is the storm drain data for Oleander. The data includes a pipeline database with a field named PipeSize and a fixtures database with a field named Type, which identifies the fixture type. These fields will be used for the summaries of each outfall. Both datasets have a field named System, into which you will store the system name that has been assigned to the metering station. The fixtures dataset also has a field named PipeSize, in which you can store the size of pipe that each fixture is connected to.

SCRIPTING TECHNIQUES

You have used all the techniques required for this tutorial before, but here is some help on the pseudo code.

The user will select an outfall before running the script. As an extra data integrity rule, make sure only one feature is selected before you continue. An if statement can check this condition.

Start selecting the lines connected to the outfall. The first selection will be to get the line connected to the outfall, and the next selection will be to select the other lines connected to the first selected line. This process will repeat until all the lines in that particular watershed system have been selected. How will you know when to stop selecting? Try setting up a count of the selected features, and then use a while statement to see whether the number has increased after an iteration of the selection process. If the number has increased, keep on selecting. Once the number remains the same through an iteration, you are done.

The selection of the attached fixtures is straightforward and can be done in a single command. Then you can use a cursor to go through the selected set of features one by one and get the pipe size from the attached line and transfer it to the point.

The final processes of storing the system name and performing the summaries can be done with a single step for each process.

This task sounds simple, but it requires some careful coding to make sure all the interrelated steps occur in the correct order. As with the other tutorials, perform it in steps, and code and debug each individual process before adding the next process. Be sure to write detailed pseudo code! Note that there is no sample pseudo code given for this script.

Combine loops

1. Open the map document Tutorial 2-8. An overall view of the storm drain system at the north end of Oleander is displayed. The purple triangles represent the outfalls for the study, as shown in the graphic. Open bookmark LBC-12A to zoom to the first outfall. From this point, you can get an idea of what a single drainage shed system looks like. Make a manual selection of the outfall, and then use the other selection tools to select all the lines and fixtures associated with this drainage shed. Also investigate the attribute tables of these feature classes, and look for the fields you will be working with. This inspection will give you an idea of what the script will be doing.

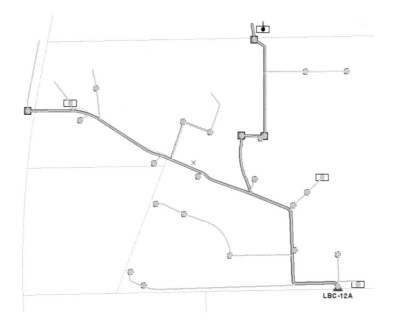

2. For simplicity, start with a simpler drainage shed, so move to bookmark WC-2B, as shown in the graphic. Once you have the script written and debugged, move on to more complex drainage sheds.

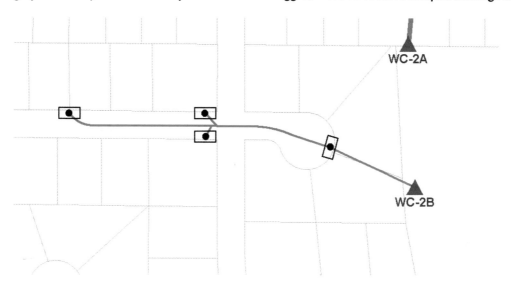

3.   After you have familiarized yourself with the process and written your pseudo code, begin writing the script. Open your IDE, and create a new script file named DrainageshedAnalysis.py with the standard lines to import module and set the working environment.

The selection process is at the heart of this script, so you should write and debug that part of the code first and create the script tool. Then, as you continue to edit the script, the script tool will gain more functionality. Start your code by making sure that only one feature is selected.

4.   Write the step to make sure that only one feature in the Fixtures feature class is selected. Otherwise, send a message to the user to let them know that the script will not run, as shown:

```
# Check to make sure that only one feature is selected before continuing
fixCount = int(arcpy.GetCount_management("Fixtures").getOutput(0))
arcpy.AddWarning("The number of features selected is " + str(fixCount) + ".")
if fixCount <> 1:
    arcpy.AddError("The number of selected features MUST be only one." +\
    " Prepare a new selection and try again.")
    raise exception
else:
    arcpy.AddWarning("Only one feature is selected and the script is continuing ...")
```

Next, get the drainage system name. You could prompt the user to supply it, or you could retrieve it from the field System in the Fixtures feature class. Determining the drainage system name sounds like a simple process, but you must set up a cursor and use the row object with the getValue object to get the value of the single feature.

5.   Set up a cursor using the Fixtures feature class. Then set up a variable to accept the value of the field System, as shown:

```
# Set up a cursor to iterate through the selected row of Fixtures
# arcpy.da.SearchCursor(in_table, field_names, {where_clause},
#            {spatial_reference}, {explode_to_points}, {sql_clause})
# By only specifying one field name, the value of row[0] contains that field value
fixCursor = arcpy.da.SearchCursor("Fixtures","System")
for row in fixCursor:
    systemName = row[0]
    arcpy.AddWarning("You are working on the " + systemName + " system.")
```

You will be making selections of both the storm drain lines and the fixtures, so make a feature layer for each of the feature classes. Remember that making selections inside a script can only be done on feature layers. An interesting note is that if you made the fixtures feature layer using the feature class in the current table of contents, the layer would contain only one feature because it has a selected feature. Instead, you should make the feature layer from the data in the geodatabase.

6.   Add the code to make a feature layer for the line and point feature classes, as shown:

```
# Selections can only be done on feature layers,
#     so create one for the lines ...
stormLinesLayer = arcpy.MakeFeatureLayer_management("MainLat")
# ... and one for the fixtures
fixturesLayer = arcpy.MakeFeatureLayer_management(\
r"C:\EsriPress\GISTPython\Data\StormDrainUtility.gdb\Storm_Drains\Fixtures")
# Note that the feature layer was made by referencing the data's location
#           in the geodatabase and not in the map document
# Referencing the data in the map document would only get the currently
#           selected features
```

7.   Add code, as shown in the graphic, to use the selected outfall to select the line(s) that intersect it. Note that this process does not use the feature layer, but rather the feature class from the table of contents with the selected feature.

```
# Use the selected outfall to select the first line(s) connected to it
# Selection type will be "Create new selection"
# SelectLayerByLocation_management (in_layer, {overlap_type}, {select_features},
#          {search_distance}, {selection_type})
arcpy.SelectLayerByLocation_management(stormLinesLayer, \
"INTERSECT","Fixtures","","NEW_SELECTION")
```

This code selects the first line (or in some cases, more than one line), which can now be used to select other lines. You can select lines that intersect this line and repeat until you have selected all the lines in this watershed system, or basically select until the count of selected features no longer increases. To accomplish this task, you can set up two count variables to hold the current count of selected features and the new count of selected features. When these two counts are equal, it means that the selection process did not select any new features and is therefore complete.

8.   Add the code to set up two count variables to hold the feature counts, as shown:

```
# With the first line feature selected, use it to select the other lines
# Set up two feature count variables
# The first will be the current selection
lineCount1 = int(arcpy.GetCount_management(stormLinesLayer).getOutput(0))
# The second will be the new selection (initialize outside the while loop)
lineCount2 = 0
```

9.   Add the while statement and the code to select and check the count of new features, as shown:

```
# Set up a while statement - it will end when the current selection equals the new selection
while lineCount1 <> lineCount2:
    lineCount1 = lineCount2
    # Use selected features to select intersecting features
    # Selection type will be "Add to selection"
    arcpy.SelectLayerByLocation_management(stormLinesLayer, "INTERSECT",stormLinesLayer,\
    "","ADD_TO_SELECTION")
    # Get count and match against previous count to see if the entire
    #      drainage shed system is selected
    lineCount2 = int(arcpy.GetCount_management(stormLinesLayer).getOutput(0))
    # Message to keep track of selections
    arcpy.AddWarning("previous set = " + str(lineCount1) + " and new set = " + str(lineCount2))
```

The while statement checks its condition before it runs. Inside the statement, set the first count variable to the previous count and the second count variable to the new count. When the statement repeats, it checks the condition again. A message line is included so that the user can check the results on the fly. This line is useful for debugging, too. As an option, you may want to include a message to signify that the selection process is finished and to report how many features are in the selected set, as shown:

```
# Message to show end of line selections
arcpy.AddWarning("Finished selecting: " + str(lineCount1) + " = " + str(lineCount2))
```

To complete the selection process, select the fixtures that intersect the selected line features.

10. Add the code to select the point features based on their intersection with the selected line features, as shown:

```
# Select only the fixtures that intersect the selected line features
arcpy.SelectLayerByLocation_management(fixturesLayer,"INTERSECT",stormLinesLayer,\
    "","NEW_SELECTION")
fixCount = int(arcpy.GetCount_management(fixturesLayer).getOutput(0))
arcpy.AddWarning("The count of fixtures is " + str(fixCount) + ".")
```

The selection processes are all complete, and you can go ahead and create the script tool and test it.

11. Create a new toolbox in your MyExercises folder named Engineering, and add a new script tool using the DrainageshedAnalysis.py script, as shown in the graphic. The script will have no input parameters. When this is done, select the outfall for system WC-2B, and run the script. You may also want to try running it without any features selected to test the count-checking routine.

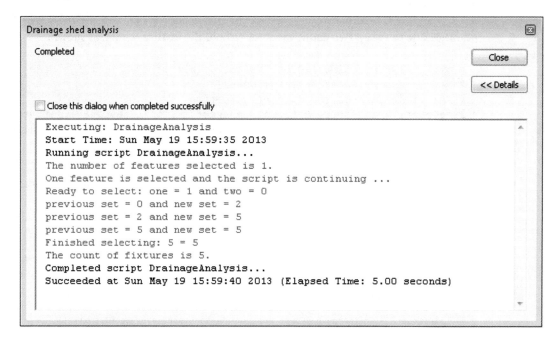

If everything checks out, continue with the script. Otherwise, finish debugging the selection processes.

The next set of processes to code is the summary statistics. For the lines, the storm water supervisor wants a table showing a summary of length categorized by line size. For the points, the supervisor wants a count of features categorized by type. If the summaries are done on the feature layers, they will contain information from only the selected features, so the code for the summaries should follow the selection of the fixtures. The names can include the system name you extracted earlier followed by either _StormLineSummary or _StormFixtureSummary.

12. **Add the code for the Summary Statistics tool for both the lines and the fixtures, as shown:**

```
# Perform a summary statistics process on the storm drain lines
# Statistics_analysis (in_table, out_table, statistics_fields, {case_field})
arcpy.Statistics_analysis(stormLinesLayer,r"C:\EsriPress\GISTPython\MyExercises\MyAnswers.gdb\\" + \
systemName.replace("-","_") + "_StormLineSummary", [["Shape_Length", "SUM"]], "PipeSize")

# Perform a summary statistics process on the fixtures
arcpy.Statistics_analysis(fixturesLayer,r"C:\EsriPress\GISTPython\MyExercises\MyAnswers.gdb\\" + \
systemName.replace("-","_") + "_StormFixtureSummary", [["Type","COUNT"]],"Type")
```

This procedure fills the requirements of the Public Works Department, and the department can use the two output tables for its calculations.

Exercise 2-8

The city has done an elevation study of the creeks in Oleander and wants to do a more detailed study in 3D using this new data, which can be found at City of Oleander > Planimetrics > AnalysisCreeks. Each segment has a field named Slope, which contains the creek's measured percentage of slope. The elevation of the outfalls (where the smaller creeks enter the major waterways) has also been recorded for each creek network, but for 3D analysis, each creek line segment must have the input and output elevations recorded. The fields for the input and output already exist as FlowLine_In and FlowLine_Out, respectively. Only FlowLine_Out for the outfall segments is recorded, which is shown in blue in the map document. Write a script to populate the rest of the flow line fields.

The user will select one of the creek outfalls and run the script tool you create. The tool will trace the creek line by line, calculate each new elevation upstream, record it, and then move on.

The key to this process is to calculate the change in elevation from one end of the line to the other. The change in elevation is found by multiplying the slope by the segment length and dividing by 100 (because the slopes are percentages). The change in elevation can then be added to the FlowLine_Out value to get the next FlowLine_In value. Repeat this step for each segment, moving upstream and using the last calculated value of FlowLine_In as the FlowLine_Out value for the next segment.

(**Hint:** This process will work until you hit a fork in the creek. Why is this a problem? What can be done to fix it?)

Use the map document Exercise 2-8 for this project.

Tutorial 2-8 review

All the techniques used in these processes were used in previous tutorials and exercises, except in this tutorial no pseudo code was given for the script. By now, you probably appreciate the importance of pseudo code in keeping track of all the steps the code must complete and the order in which they must occur. If you completed this tutorial successfully, designing and coding all the processes without the aid of prewritten pseudo code, you are well on your way to becoming a Python programmer.

Study questions

1. What was the hardest part about writing pseudo code for this project?
2. Does it help or hurt to add messages in the code to report progress?
3. The potential for an endless loop existed in this project. Can you explain what it was and how you mitigated it?

Tutorial 2-9 Creating custom toolbars

Script tools make running your tasks from a custom toolbox faster and easier, but these tools may also be placed on a custom toolbar, making them more convenient.

Learning objectives

- Creating custom toolbars
- Adding tools and scripts to custom toolbars

Preparation

Research the following topics in ArcGIS for Desktop Help:

- "Creating a new toolbar"
- "Adding and removing tools on menus and toolbars"

Introduction

Toolbars are easy to create and can contain your custom script tools, models, and system tools. You can categorize items on a toolbar, and even make drop-down submenus. It is important to note, however, that the tools still operate as stand-alone tasks and do not interact with each other. Also, any custom toolbars you create through the customized interface exist only in the map document in which they are created. These toolbars cannot be shared easily with other users.

Scenario

The script tool you created in tutorial 2-8 requires that an item be selected before the tool runs. It would make things easier for the user if you created a separate toolbar and assembled all the tools together to accomplish the task, including the script tool, the Select Features tool, and the Clear Selected Features tool. As an option, you can include the Continuous Pan tool to make moving around the map area easier.

Data

Use the script tool you created in tutorial 2-8, along with several system tools.

SCRIPTING TECHNIQUES

Toolbars are created in the Customize Mode and accessed from the Customize menu, which makes any toolbar customizable.

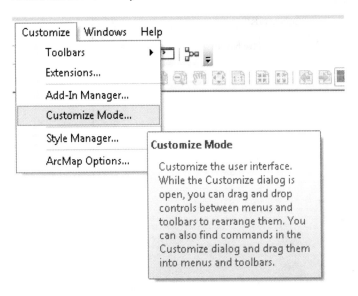

When you click New in the Customize dialog box, as shown, a new toolbar is created.

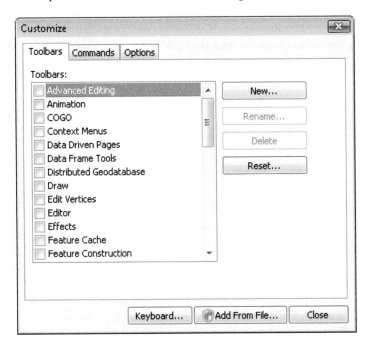

Then you can simply drag your tools to the toolbar. To make custom drop-down menus on a toolbar, go to the Commands tab, scroll all the way to the bottom, and select New Menu, as shown in the graphic. This menu can be dragged to the toolbar and customized.

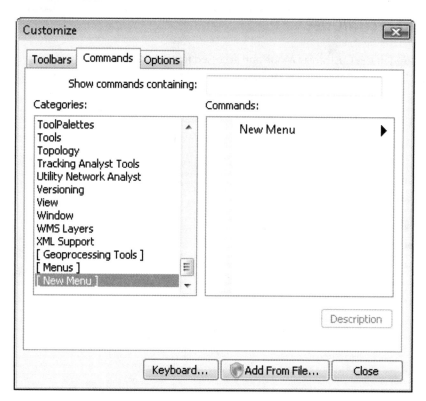

Once tools are added to the toolbar, the toolbar will perform exactly like a system toolbar. You can drag the customized toolbar anywhere on your screen or dock it to any of the existing menus. Remember, however, that the toolbar does not exist beyond the map document in which it is created.

The sample in the graphic shows a variety of tools and menus on a custom toolbar.

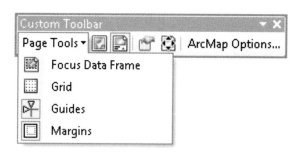

Create a custom toolbar

1. **Open the map document Tutorial 2-9. On the main menu, click Customize > Customize Mode.**

Note all the existing toolbars listed in the Toolbars panel. Any of the toolbars that have a check mark next to them are currently visible in the map document. Notice also that every tool on every toolbar appears as active, meaning that each toolbar can be moved or deleted as desired.

2. **Click New at the right of the Toolbars panel, and name the new toolbar Storm Drain Tools, as shown. Click OK.**

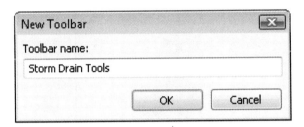

The toolbar is created, and by default it will be added to the map document. The toolbar may be docked in the menu area at the top of your map document, or it may be free-floating in the map area, as shown:

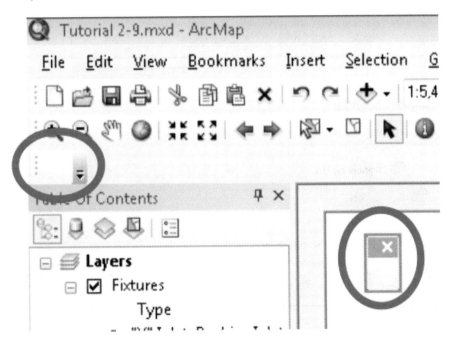

3. **Undock the toolbar, and drag it just above the Customize dialog box in the map area to make it easier to work with.**

The script tool you created must be brought into the Customize interface before it is available to use on menus.

4. On the Customize dialog box, click the Commands tab, and scroll to the bottom. Click [Geoprocessing Tools] and click the Add Tools button, as shown:

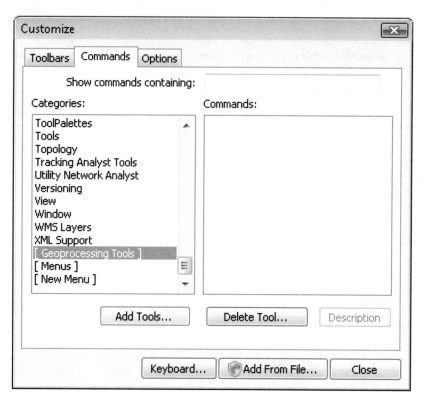

5. When prompted, navigate to the Engineering toolbox, and select the Storm Drain Watersheds tool, as shown. Click Add.

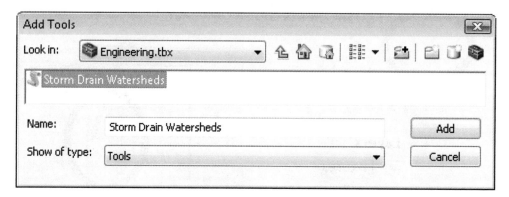

The tool is now available to be placed on any toolbar simply by dragging it to the toolbar.

6. Drag the Storm Drain Watersheds tool to your new toolbar, and drop it there.

As shown in the graphic, the tool appears on the toolbar, but there is still some customizing to do. You can control the icon that is displayed, the text that is displayed, and what will be shown to the user.

7. Right-click the new tool on the toolbar. Note the options displayed. Change the name to Trace Watersheds and click Text Only, as shown. This will display only the name of the tool.

In addition to the storm drain tool, you also need to add a new drop-down menu with some feature selection tools.

8. On the Customize dialog box, click [New Menu]. In the Commands panel, click *New Menu* and drag it to the left side of your toolbar, as shown:

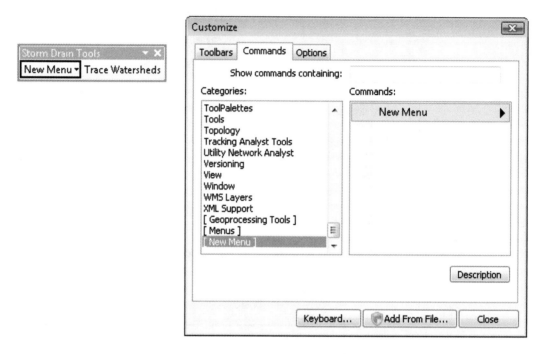

Notice that the menu item has a drop-down arrow, which indicates to users that it is a drop-down menu rather than an active tool.

9. Right-click the menu item, change the name to Selection Tools, as shown, and press Enter.

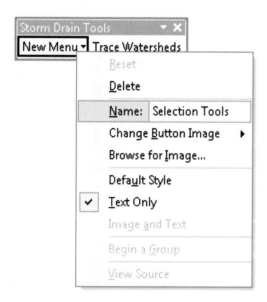

This new menu will hold the Select By Rectangle tool and the Clear Selected Features tool. You can use the Search feature on the Customize dialog box to find any tool in any category.

10. On the "Show commands containing: " line, type Select By. **Look down the resulting list and find Select by Rectangle. Then drag it to the toolbar, placing it below the drop-down menu, as shown:**

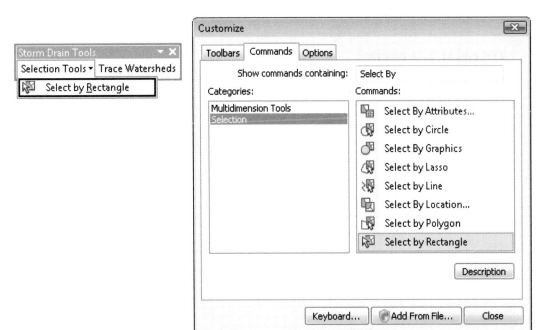

Your turn

Use the Search feature to locate the Clear Selected Features tool, and add it to the drop-down menu. Then locate the Continuous Pan tool, and add it to the right side of the toolbar. When you are done, close the Customize dialog box. Examine the toolbar and the tools it contains. Note that the Clear Selected Features tool appears dimmed, as shown in the graphic, because features must be selected for it to become active.

The toolbar is now ready for use and contains a combination of custom tools that you created and system tools from the ArcGIS toolboxes.

Exercise 2-9

In exercise 2-7, you wrote some custom script tools for the Fire Department, and you feel that the department might also benefit from having its custom tools accessible from a single location.

Open the map document Exercise 2-9. Create a custom toolbar to contain the script tools you created for the Oleander Fire Department in exercise 2-7.

Tutorial 2-9 review

The creation of custom toolbars will make your applications appear even more professional. You can include all the custom script tools you have created and system tools. The toolbars you create can also be docked or kept free floating. They cannot, however, be shared with others. These types of toolbars are saved only to your personal files and are accessible only through your map documents.

Study questions

1. Why is it useful to add existing system tools to your custom toolbar?

2. Can you control the availability of a tool on a custom toolbar—in other words, can you cause a tool to be disabled until certain conditions are met, such as when an item is selected or when you are in data view?

3. When would you use a menu, and when would you use a toolbar?

Chapter 3

The ArcPy mapping module

Introduction

The ArcPy module includes a special set of tools for working with your map documents called the *ArcPy mapping (arcpy.mapping) module*. This module provides access to all the elements of a map document and allows you to control the output for custom maps and map books. The tutorials in this chapter demonstrate how to access map documents, data frames, graphic elements, and data layers. Each of these elements has a unique set of properties for controlling these elements.

Tutorial 3-1 Accessing map document elements

Working with the graphic elements on the map page is one of the most basic tasks of map creation, but also one of the most powerful. Each element can be altered with Python code you write to create maps and map books.

Learning objectives

- Controlling the map document elements
- Listing and controlling map layers
- Accessing features in a list object
- Setting input filters and default parameter values

Preparation

Research the following topics in ArcGIS for Desktop Help:

- "MapDocument (arcpy.mapping)"
- "ListLayoutElements (arcpy.mapping)"
- "ListLayers (arcpy.mapping)"
- "Customizing script tool behavior"

Introduction

If you are familiar with geoprocessing in ArcMap and ArcCatalog, you are familiar with most of the tools that you would use in stand-alone Python scripts and the script tools available with the ArcPy mapping module. However, the mapping module introduces a new concept of working with the elements within map documents. The module allows you to work with data frame parameters, components of the table of contents, and characteristics of elements of a currently open map document in ArcMap or with those in a saved map document.

One standard line of code is all that is necessary to access a map document and gain control of it. The code to access a map document that is currently open in ArcMap looks like this:

```
arcpy.mapping.MapDocument("CURRENT")
```

The code to access any map document without having to open it looks like this:

```
arcpy.mapping.MapDocument(r"c:\EsriPress\Maps\Tutorial.mxd")
```

You can set the map document equal to an object name, which provides access to all the current map document parameters in ArcMap, or replace the keyword CURRENT with the name of another map document to access its parameters without opening it. These parameters include classes, such as identifying the data frames and noting which one is active; controlling the map documentation, such as author, description, tags, and title; and characteristics, such as the page size and file path.

In addition to the document's characteristics, you can manipulate layers in the table of contents. Functions in the mapping module allow you to add or alter layers in the table of contents, which may include something as simple as handling their visibility to something as complex as reworking their symbology.

A set of list functions also exists. These list functions are much like the list of ArcPy functions you used in chapter two that let you build list objects of things such as bookmarks, layers, data frames, and layout elements. The key to using list functions, however, is that each item placed in your map must be assigned an alias. The list functions all access the items based on their name or alias. An interesting thing to note about the list functions is that a list object is always returned, even if the results have only one value, which will affect how you retrieve and manipulate the values.

ArcGIS for Desktop Help has a list of all the classes and functions available in arcpy.mapping and provides good examples, tool descriptions, and sample code to help you understand how these tools can be used.

Scenario

The city planner has asked that you show the parcel data in several maps: one map to show the land use, another to show the appraised value of the properties, and the third to show the age of the structures grouped into 10-year categories.

With arcpy.mapping, you can use a basic map template to change the components necessary to create each of these maps. The nice thing about using a template is that you do not have to worry about opening ArcMap and finding the appropriate layers, and all three maps will have a similar look and feel. You can even export the results to a PDF document that can be e-mailed to the city planner. The supplied template has symbolized map layers in the three required classifications. Determine which map type is selected from the user input window, and make the correct layers visible. Then export a completed map.

Data

A map template named ParcelTemplate is ready for this project. This template holds the parcel data that you need, is symbolized in the three required methods, and has a map layout with titles and descriptions that you can change for each output map.

SCRIPTING TECHNIQUES

You will find it easy to manipulate elements within your map using the arcpy.mapping module in your scripts, and there are three basic techniques you need to perform in these scripts. The first of these techniques is to create a map document object that references the map document you wish to work with. The introduction to this tutorial shows how to create this object using either the keyword CURRENT or the path name of a stored map document.

The second technique is to create a list object of all the data frames in the map document you have selected. This is done with the ListDataFrames() function. Each data frame is added to the list object, and the data frames are accessed by using the object's name and the index number of the data frame. Remember that the index numbers start at zero, so if there is only one data frame in your specified map document it will have the index number zero. You will learn how to write the code for this technique in this tutorial.

The final technique for working with the arcpy.mapping module is to access the elements in your map document. The technique requires that you first assign a name to each map element. For instance, if you wanted to work with a title and subtitle, you would need to open the properties for each text item and assign them a name such as mainTitle and mapSubTitle. The next step is to create a list object of the map's elements using the ListLayoutElements() function. The list will contain the names you have assigned to each element, making it easy to reference elements by name, such as mainTitle or mapSubTitle. As with other list elements you have created and worked with, you can use a for statement to iterate through the list, and then use an if statement to find the specific element of interest.

The script in this tutorial will totally automate the creation of three maps as well as alter the map elements for each unique map. There are three preset names and descriptions for the three types of maps you will create. In the script tool dialog box, you will set up a menu for the user with a value list, and then use the tool validation code to totally automate the text entry for the various maps. Although this example script shows tool validation code being set up to control map elements, it could also be used to control any aspect of the user's input for a script.

Access map document elements

1. **Start ArcMap, and open the map document ParcelTemplate.**

The template looks as if it will make fine maps, but you must do some preliminary setup work to be able to access all the map elements through a script. Each element in the map must have a unique name set as a property. When you search through map elements with a list command, you need to be able to uniquely identify each map element and understand what each element represents. It is also important to make sure that there is an existing map element for everything you might want to work with on the map. The mapping module works only on existing map elements and does not have the capability to add new elements.

2. **In the table of contents, right-click the data frame name, and open the properties. Click the Size and Position tab. Notice that the element name is the default name Layers. Rename the element** Parcel View, **as shown, and close the Properties window.**

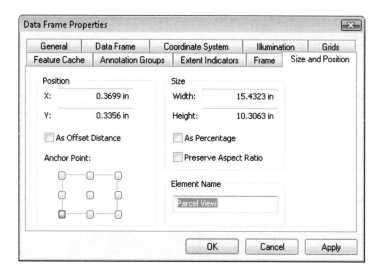

If there were more than one data frame in the map, you would need to give them all a unique and descriptive name. Next, create names and set the properties for the other map elements.

3. **In the map layout, right-click the map title text and click Properties. Note that the title has no element name. Name the element** Map Title, **and close the Properties window.**

Notice that the anchor point of the text is the lower-left corner. As you change the title, and the length of the text becomes longer or shorter, the text grows away from or toward this location, which is important to note because it can ensure that your text stays in the map area when you change it. As you look at the other map elements, make sure their anchor points will allow them to change without running off the edge of the map page. It is also good to examine the text justification on the Text tab to make sure that the anchor point matches the text justification. For example, if you use the lower-left corner for the anchor point, make sure the text is left justified; or if you use the right corner, make sure the text is right justified. The anchor point for text represents the location relative to the unrotated text—lower left is relative to the text as it is read, not relative to its appearance on the map.

The map layout has several other elements, some of which you will not change in your map creation. However, each element needs to have a unique and recognizable name so that you can identify which ones you will be changing and, just as importantly, which ones you will not be changing.

Your turn

Open the properties of each of the map elements, and give them the element name shown. Also, set the anchor point and text justification to an appropriate setting.

City logo image: **Oleander Logo**

Text box at left of map: **Map Description**

Legend: **Legend** (**Hint:** *check the anchor point for the legend.*)

Map subtitle: **Map Subtitle**

City of Oleander: **City Name**

Date: **Map Date**

North arrow: **North Arrow**

Because all the elements in the map are assigned a name, you will have no trouble identifying each one correctly in the script, which does not have your WYSIWYG instincts.

Begin to build the script to manipulate the map. For this tutorial, build and run the script in ArcMap so that you can see the changes to the map as you go along. Create a new script, make it a script tool, and then run and edit it in the Catalog window in ArcMap.

4. **Write pseudo code for this script, including changing the map elements and layer symbology.**

5. **Start your IDE, and create a new Python script named** ThreeWayMap.py. **Without adding any code, save and close the script.**

6. In the Catalog window, navigate to the Custom Python Tools toolbox in your MyExercises folder. Right-click the toolbox and click Add > Script. Give the script a name, label, and description, as shown:

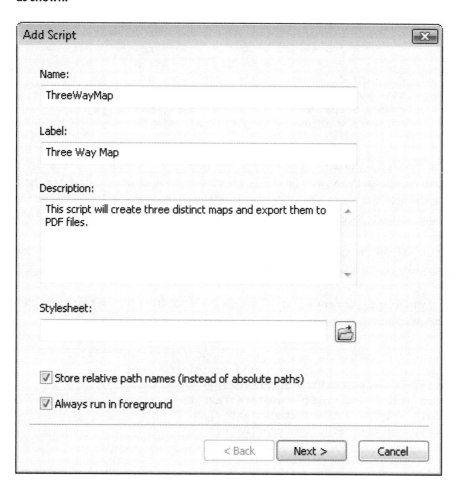

7. Set the tool to point to the new script you created. There are no input parameters yet, so leave that portion blank. Click Finish to create the script tool.

8. Right-click the script tool and click Edit.

Your IDE opens, and you can make changes to the Python code and have the script tool recognize the changes as soon as you save them. **Note:** if your IDE of choice does not open, see appendix A for instructions on how to set the default IDE for ArcGIS.

Because you are running this script in ArcMap, you do not need to import ArcPy or set any environments. Everything you will do is associated with this map document, but there are certain lines of code that you will use in all your mapping scripts that you can add to a template for future use. These lines include commands to get the current map document, the data frame, and a list of the layers in the data frame. For these commands, use the functions MapDocument(), ListDataFrames(), and ListLayers(). Look up these functions in ArcGIS for Desktop Help to determine their syntax and to see code samples.

9. Add the code to create a map document object named thisMap using the CURRENT keyword. Then create a data frame list named myDF and a list of the layers it contains, and save these lists to a list object named myLayers, **as shown:**

```
# Create a map document object for the currently open map document;
# a data frame list of the only data frame in this map document,
# Note: leave off the index [0] to get all data frames;
# a list of layers in this map document
thisMap = arcpy.mapping.MapDocument("CURRENT")
myDF = arcpy.mapping.ListDataFrames(thisMap)[0]
myLayers = arcpy.mapping.ListLayers(myDF)
```

Note that this code is written specifically for a map document with one data frame. The inclusion of the zero-numbered index value (index[0]), means that it will automatically define the data frame object as the first data frame. The list of layers contains all the layers in the map document, with the layer at the top being indexed at zero and increasing by one as you go down the table of contents.

Prompt the user for the names and descriptions that you want to use in the map. Ask for a map title, a map subtitle, a current date, and a description for the text box.

10. Add four variables to your code that will accept the input from the user for map title, subtitle, date, and description, **as shown:**

```
# Create variables to accept user input for the map title,
# map subtitle, current date, and description.
newTitle = arcpy.GetParameterAsText(0)
newSubTitle = arcpy.GetParameterAsText(1)
newDate = arcpy.GetParameterAsText(2)
newDesc = arcpy.GetParameterAsText(3)
```

With this information on hand, you can begin to alter the text properties of the text type map elements. Use the ListLayoutElements() function to create an indexed list object of all the map elements. Research this function in ArcGIS for Desktop Help, and you'll find that it can be restricted to a certain element type, such as just the text or the graphic elements, and that it may include a wildcard. The elements will be indexed in the list object, but there is no way to predict in what order, so when you access them, use the names set earlier.

11. Add the code to create a list object named myElements of the text elements that start with the word *map*, **as shown. Note that the wildcard is not case sensitive.**

```
# Create a list of the map elements
# These will be accessed by their names
myElements = arcpy.mapping.ListLayoutElements(thisMap, "TEXT_ELEMENT", "map*")
```

The code is in place to update the text for each of the map elements. Once you identify each of the elements by name, replace the element's text with the user input. When the updates are made, refresh the map using the RefreshActiveView() function to see the changes.

12. Write the code to loop through each of the layout elements, and match the name with the new text entered by the user, as shown. (Hint: use a for statement to scroll through the elements and an if statement to find the element name and to decide which elements you want to modify.)

```
# Scroll through the list of elements with a for statement
# Use an if statement to test for each of the element names you
# are changing and replace the text with the user input
for element in myElements:
    if element.name == "Map Title":
        element.text = newTitle
    elif element.name == "Map Subtitle":
        element.text = newSubTitle
    elif element.name == "Map Date":
        element.text = newDate
    elif element.name == "Map Description":
        element.text = newDesc

# Refresh the map so that the changes can be seen
arcpy.RefreshActiveView()
```

The script is ready to change the appropriate map elements, but the script tool is not. When you created the tool, no user inputs were defined. Now that you have added those statements to the script, you must modify the script tool to accept the input.

You want this adjustment to be as foolproof as possible so that the city planner will not encounter any problems with the script. Because the output can be only one of three maps, you could give the user a drop-down list to select the title. Then the subtitle and map description could be a block of preset text, depending on the chosen map title. The only other information the user will need to provide is the month and year to display on the map. Because this information represents the date the data was last updated, you would not necessarily want to use today's date, but rather let the user supply it. All of this can be done with the tool validation code.

If the user selects the map to display Property Value when they run your script, set the following text for subtitle and description:

Derived from Appraisal District files

The property values for Oleander are appraised and determined by the Tarrant County Appraisal District. Monthly reports are used to update the parcel database to keep the data as current as possible. Property shown with a value of $0.00 may be under protest, and the new value will be available after the owner has had a public hearing to settle the matter.

If the user selects the map to display Land Use, set the following text for subtitle and description:

Current land use for all property

Land use is determined by city staff to fit into one of many categories. The initial land use is determined by the occupant's Certificate of Occupancy application and later verified in the field. Any changes in land use are noted on an annual basis, which may trigger the occupant to obtain a new Certificate of Occupancy or if necessary a Specific User Permit.

If the user selects the map to display Date of Construction, set the following text for subtitle and description:

Historical construction data shown by decade

The records of construction for Oleander date back to the 1940s when Oleander was first incorporated. Buildings constructed prior to that date are grouped into the "pre-1940s" category since the city does not have verification of dates earlier than that. Vacant property and property where no construction date is known are shown with no color shading. The purpose of this map is to give an overview of the growth of Oleander over the years and not to determine a specific building's date of construction.

13. **Save and close the script. Right-click the script tool, open the properties, and then click the Parameters tab. On the first line, add a display name of** Map Title **with a data type of String. Set the filter to Value List, and add the three map titles to the list, as shown:**

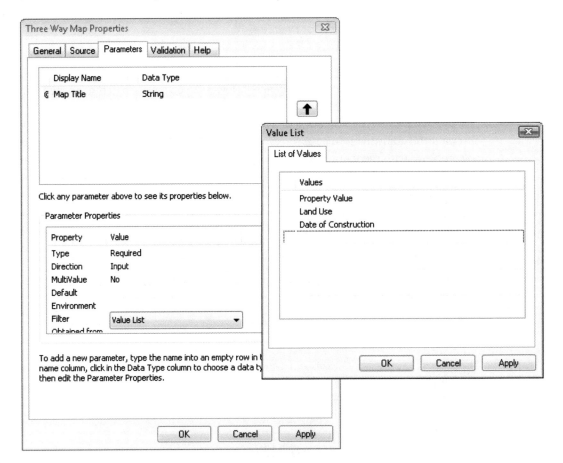

14. Enter the other three input display names as Subtitle, Date, and Description, in that order, and each with a data type of String, as shown in the graphic. Click Apply. (Hint: the order is determined by the order in which the script indexes the map elements.)

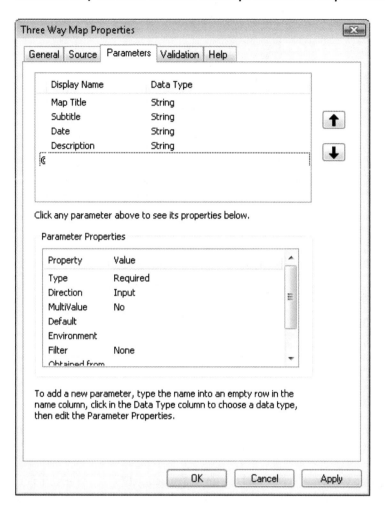

The tool validation code can be used to set any of the text values, but the user will be able to type additional information if desired.

In the tool validation code, the updateParameters class is used to accept and monitor the inputs for the script. Each value typed by the user is placed in the self object and indexed in the same order as referenced in the script. For example, the first parameter is self.params[0], and the second is self.params[1]. These objects have properties such as .enabled, which determines whether the input value has been changed, and .value, which is where the entered text is stored. Your code will check to see if the input for map title, chosen from the value list, has been changed. If a change is detected, use an if statement to determine which title was selected, and then set the subtitle and map description to the preset text.

15. Open the properties of the script tool, and click the Validation tab. Click Edit, add the code necessary to determine whether the title has been changed, and set the other parameters accordingly, as shown. Save the code edits, apply the changes, and close the Parameters window.

```python
import arcpy
class ToolValidator(object):
    """Class for validating a tool's parameter values and controlling
    the behavior of the tool's dialog."""

    def __init__(self):
        """Setup arcpy and the list of tool parameters."""
        self.params = arcpy.GetParameterInfo()

    def initializeParameters(self):
        """Refine the properties of a tool's parameters.  This method is
        called when the tool is opened."""
        return

    def updateParameters(self):
        """Modify the values and properties of parameters before internal
        validation is performed.  This method is called whenever a parameter
        has been changed."""
        if self.params[0].altered:
            if self.params[0].value == "Property Value":
                self.params[1].value = "Derived from Appraisal District files"
                self.params[3].value = "The property values for Oleander are appraised and determined by the " + \
                "Tarrant County Appraisal District. Monthly reports are used to update the parcel database to " + \
                "keep the data as current as possible. Property shown with a value of $0.00 may be under protest " + \
                "and the new value will be available after the owner has had a public hearing to settle the matter."
            elif self.params[0].value == "Land Use":
                self.params[1].value = "Current land use for all property"
                self.params[3].value = "Land use is determined by city staff to fit into one of many categories. " + \
                "The initial land use is determined by the occupant's Certificate of Occupancy application, and " + \
                "later field verified. Any changes in land use are noted on an annual basis, which may trigger " + \
                "the occupant to obtain a new Certificate of Occupancy or if necessary a Specific User Permit."
            else:
                self.params[1].value = "Historical construction data shown by decade"
                self.params[3].value = "The records of construction for Oleander date back to the 1940s when " + \
                "Oleander was first incorporated. Buildings constructed prior to that are grouped into the pre-1940s " + \
                "category since the city does not have verification of dates earlier than that. Vacant property and property " + \
                "where no construction date is known are shown with no color shading. The purpose of this map is to " + \
                "give an overview of the growth of Oleander over the years and not to determine a specific building's " + \
                "date of construction."
        return

    def updateMessages(self):
        """Modify the messages created by internal validation for each tool
        parameter.  This method is called after internal validation."""
        return
```

At this point, you have completed the code for the user interface and text element portions of the application. Test what you have completed before moving on.

16. Double-click the Three Way Map script tool to run it. Try using different selections in the dialog box, and note how the other fields change, as shown. Select one, add a month and year for the date, and click OK. Notice the changes in the map.

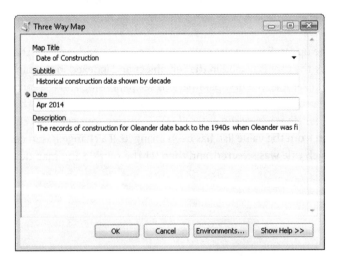

In the map template, three layers are symbolized for the output maps: Property Value, Land Use, and Date of Construction. Once you determine which title has been chosen by the user, scroll through the layers, and use the name and visible properties to turn the correct layers on (or off).

17. Add an if statement and a for statement to determine the selected title, and use that information to determine which layers to make visible, as shown:

```
# Determine which title has been chosen
# Use the title name to determine which layers to make visible
if newTitle == "Property Value":
    for layer in myLayers:
        if layer.name == "Property Value":
            layer.visible = "True"
        elif layer.name == "Land Use":
            layer.visible = "False"
        elif layer.name == "Year of Construction":
            layer.visible = "False"
        else:
            layer.visible = "True"
elif newTitle == "Land Use":
    for layer in myLayers:
        if layer.name == "Property Value":
            layer.visible = "False"
        elif layer.name == "Land Use":
            layer.visible = "True"
        elif layer.name == "Year of Construction":
            layer.visible = "False"
        else:
            layer.visible = "True"
else:
    for layer in myLayers:
        if layer.name == "Property Value":
            layer.visible = "False"
        elif layer.name == "Land Use":
            layer.visible = "False"
        elif layer.name == "Year of Construction":
            layer.visible = "True"
        else:
            layer.visible = "True"

# Refresh the map so that the changes can be seen
arcpy.RefreshActiveView()
```

Note the addition of the RefreshActiveView() function. This function refreshes the map on your screen and makes the changes visible. If you were working with a closed map document, you would not need to include this function for the changes to be made on the map.

The last requirement is to output the map to a PDF document. This can be done using the ExportToPdf() function and storing the output to a folder.

18. Add a line of code to export the current map document to a PDF document. Use the map title concatenated to the date to create the output file name, and save it to your MyExercises folder, as shown:

```
# Output completed map to a PDF file
# Use the map title and supplied date in the title
arcpy.mapping.ExportToPDF(thisMap,r"C:\EsriPress\GISTPython\MyExercises\\" + \
newTitle.replace(" ","_") + "_" + newDate.replace(" ","_"))
```

A little housekeeping is needed when working with arcpy.mapping. You should always delete the map document object when you are finished to prevent the map document from being locked to other users.

19. Add a line of code at the end of your script to delete the map document object thisMap, as shown. Save and close your script.

```
# Clear this map document from memory so that it won't be locked
del thisMap
```

Note that this does not delete the map document, but only the reference to it in your script.

20. Try running the Three Way Map script tool for different map titles, and note the results. Check that the PDF document was created successfully.

With this script completed, these monthly exhibit maps will be created quickly and easily with little user interaction. You may even turn this task over to the city planner because it will be virtually impossible to create the wrong map.

Exercise 3-1

In ArcGIS for Desktop Help, research the tools used to set the properties of the layer objects used in the Three Way Map script tool, and in particular look at the methodology of changing the characteristics of the symbology scheme. The classification of a layer can be altered, but the classification type cannot be changed.

The city planner has asked that the application be altered to allow him to select which year's values are used for the Property Value map. Add a value list to the input box that will ask which property value field should be used if the map type selected is Property Value. If any other type of map is selected, make this input inactive using the tool validation code. (**Hint:** this is done with the .enabled property of the self.params[] object.)

Once the map type of Property Value is selected, and the user identifies which field to use, produce the requested map. The other map titles will work the same as in tutorial 3-1.

Tutorial 3-1 review

This tutorial uses two types of lists: one to get a list of the elements in the map document, and one to get a list of the layers. As you have learned, both lists are controlled with indexes. Another list was used to list the data frames in the current map document, but it returned only one value. Because there was only one data frame listed, by default, it has an index number of zero, so code was added to make the returned value a single data frame object rather than a data frame list. After this addition, any references to the data frame do not need to include the index number. The following code gives an example of each method of acquiring a layer object for processing and shows one way to make your code more concise:

```
# Adding the [0] to the end of the list statement when there
# is only one returned value makes the output a data frame object
# rather than a list object
myDF = arcpy.mapping.ListDataFrames(thisMap)[0]

# Other examples would be to get a single layer with a wildcard
# and not have to search through a list for a particular layer
creekLayer = arcpy.mapping.ListLayers(thisMap,"Creeks",myDF)[0]

# or find the layer with a for statement
myLayers = arcpy.mapping.ListLayers(thisMap,"",myDF)
for layer in myLayers:
    if layer.name == "Creeks":
        # perform analysis
```

The basis of this tutorial is to turn layers on and off. As you learned, the layer property for visibility is easily manipulated. There are many other properties you can change, including name and transparency.

You also learned another use of the tool validation code. In this case, it is not checking to see if the values the user entered were valid, but taking other actions based on the user's input. As a data integrity technique, the validation code can prevent a lot of misspellings and incorrect data entry.

Study questions

1. What changes could you make to the script for tutorial 3-1 so that you could run it without opening a map document?
2. Where else might you have used the indexing technique to return a layer object rather than a list object?
3. Where else might you use the data validation technique used in this tutorial?

Tutorial 3-2 Controlling the map extent

You have learned how to interact with the elements of a map document using the ArcPy mapping module by changing the parameters of various items. There are many more characteristics of the map document that you can interact with, which you will explore in this tutorial.

Learning objectives

- Accessing layers in the table of contents
- Altering the map extent
- Exporting a map to a PDF document

Preparation

Research the following topics in ArcGIS for Desktop Help:

- "DataFrame (arcpy.mappping)": getSelectedExtent()
- "DataFrame (arcpy.mappping)": panToExtent()

Introduction

Many elements of the map can be stored as an object and manipulated using the Describe() and List() functions. These elements include properties of the map document itself, such as the description, title, and tags, and methods to create a thumbnail image or save the map document in the current and earlier version formats. Access to these functions is through the map document object, which you learned to create in tutorial 3-1, and which is created in every script that uses arcpy.mapping.

Manipulating these items is one thing, but the real power of the mapping module comes in being able to manipulate the data within the map and in how the data is displayed in the data frame. You can change the scale of the map or zoom and pan the map, and you can also add and control data layers within the map. The mapping module lets you add data, change the symbology, select a scale and spatial reference, and even change the map extent to match selected data layers. A good tutorial included in ArcGIS for Desktop Help is the topic "Creating Data Driven Pages." Data Driven Pages basically works as a map book generator.

In this tutorial, you will develop an application that changes the page elements and exports the results to a PDF document.

Scenario

The City of Oleander receives frequent requests from real estate agents for small exhibit maps of properties they are showing. The agents typically e-mail the tax account number to identify the subject tract. As an aid to creating these maps, design a tool that will create all the maps once the parcel is selected. The agents will be able to choose a parcel map showing the property with subdivision names and lot numbers, a water utility map, a sewer utility map, or a storm water utility map. The utility maps will also show the building footprints and the addresses.

Data

A template drawing has already been created, so your application will only need to accept input from the user, select and pan to the target feature, turn on or off the appropriate layers, set the map titles, and export the results to a PDF document. The Parcels layer has a field named EKEY that contains the tax account number.

The template has six group layers that you will turn on or off to make the maps. The user will be able to choose from four maps, and the group layers to show with each map are noted here:

- Parcel Map—turn on the groups Parcels Group and Base Group
- Storm Water Utility Map—turn on the groups Storm Water Utility Group, Physical Features Group, and Base Group
- Water Utility Map—turn on the groups Water Utility Group, Physical Features Group, and Base Group
- Sewer Utility Map—turn on the groups Sewer Utility Group, Physical Features Group, and Base Group

SCRIPTING TECHNIQUES

In tutorial 3-1, you learned how to create a map document object, a data frame object, and a layer list object. These objects are created in every script in which you want to gain control of the map document and all of its associated elements and properties.

You also learned how to create a filter and a value list for input. Perform these tasks again to present the user with a list of maps to choose from. Then determine which maps were chosen, and use that information to configure the display and generate the outputs.

Use the input from the user to find the subject tract in the Parcels layer. Previous tutorials showed how to make a feature layer and use that layer to select features. This is the required technique when you are running a script outside an ArcMap edit session. When running a script in an ArcMap edit session, you access the layer from the table of contents. The layer list object that you create will hold the Parcels layer, and you can find and use that layer for the selection.

Once you make the selection, pan the map to the selected feature. To get the extent of the selected feature, use arcpy.mapping to access the getSelectedExtent() method. This extent can then be fed into a method of the data frame named panToExtent(), which moves the map to the subject tract and maintains a scale of 1" = 100'. If you wanted to change the zoom and the display locations, you could use the zoomToSelectedFeatures() method, which changes the scale property of the data frame. In this tutorial, however, the scale is fixed.

Control the map extent

1.   **Start ArcMap, and open the map document Tutorial 3-2, which is shown in the graphic:**

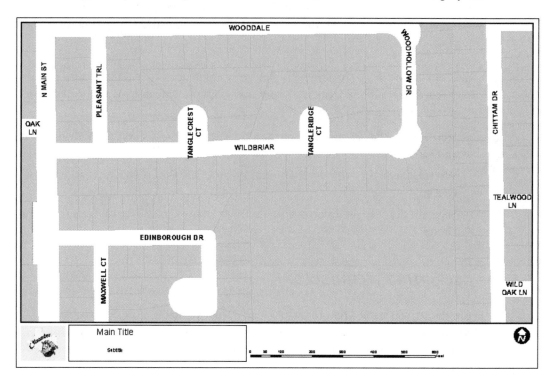

This map template will be used to create the exhibit maps for the real estate agents. First, however, make sure the graphic elements in the map have appropriate names so that you can access and control them.

2.   **Select each graphic element in the map layout, open its properties, and set an appropriate name, as shown in the list. Also, change the location and anchor point of the title and subtitle elements to accommodate whatever text is provided when the maps are created.**

Data frame = **Exhibit**
Main title = **Main Title**
Subtitle = **SubTitle**
City logo = **Logo**
Title box = **Title Box**
Scale bar = **Scale Bar**
North arrow = **North Arrow**

It is important to give every element in the map a unique name so that names are not repeated and values are not null.

Write your pseudo code, and design the steps of accepting user input and manipulating the map. Sample pseudo code for this script is shown at the end of this tutorial.

3. **Create a new Python script, and name it** RealtorExhibit.py. **Add the standard code to import the ArcPy module.**

The first part of the script is to accept the user input. Because this is a script tool in a toolbox, use the getParameterAsText() function for the input. Later, you will create the input box to accept a list of the maps the real estate agent wants.

You will also accept the text for the main title and the subtitle and the tax account number of the subject tract from the user.

4. **Add the code to accept user input into a variable named** mapList. **Add the code to accept a main title, a subtitle, and a tax account number, as shown. Save and close the script when you are finished.**

```
# Prompt the user for the maps to draw.
mapList = arcpy.GetParameterAsText(0)
# Prompt the user for a main title
mainTitle = arcpy.GetParameterAsText(1)
# Prompt the user for a subtitle
subTitle = arcpy.GetParameterAsText(2)
# Prompt for a single tax account number
taxAccount = arcpy.GetParameterAsText(3)
```

Create the script tool, and set up the user input interface.

5. **In your MyExercises folder, create a new toolbox named** Tutorial 3-2. **Start the process to add a script, and give the script the name and description shown in the graphic. Click Next.**

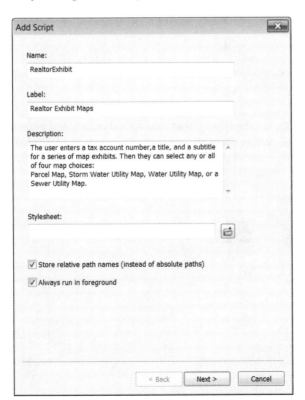

Add Script

Name:

RealtorExhibit

Label:

Realtor Exhibit Maps

Description:

The user enters a tax account number, a title, and a subtitle for a series of map exhibits. Then they can select any or all of four map choices:
Parcel Map, Storm Water Utility Map, Water Utility Map, or a Sewer Utility Map.

Stylesheet:

☑ Store relative path names (instead of absolute paths)
☑ Always run in foreground

< Back   Next >   Cancel

6.   Navigate to and select the script RealtorExhibit.py, as shown in the graphic. Click Next.

The next screen will accept the setup for the inputs. Remember that these must go in order of the index numbers you assigned in the script. The user will have the option to select any or all of the preset map types, but you will need to build an input dialog box to accommodate this selection. The technique is to make the input a data type of String with the MultiValue parameter set to Yes. Then provide a value list in the filter dialog box. When the script runs, the user will be presented with a set of check boxes, and each value that is checked will appear in the variable created. Note that the selected values will be in a regular text variable, separated by semicolons, and will not have the proper structure to be a list object.

7.   On the first line, enter a display name of Select the map(s) to create, and set the data type to String, as shown:

8.   Set the MultiValue parameter to Yes, set the Filter parameter to Value List, and enter the list of map names, as shown in the graphic. When these are entered, click Apply and then click OK.

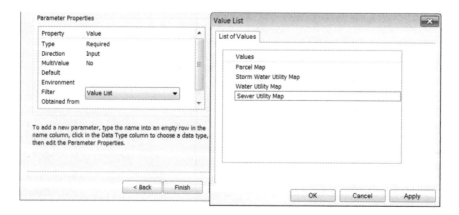

Your turn

Add the display name and data type to accept the other three variables in this script, as shown in the graphic. Make the variables all strings, and transform them into other data types as necessary within the script. When you have entered all the variables, click Finish to create the script tool.

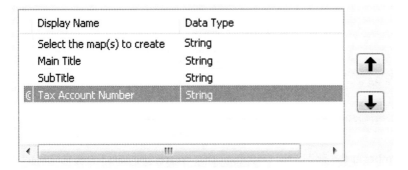

9. Test the input interface by double-clicking the script tool. The following graphic gives you an idea of what the users will see when they run the script. When you have finished your investigation, click Cancel.

With the data entry part of the script working, you can focus on the feature selection tools. The script will need to select the subject tract based on the number the user entered, in the same manner as previous tutorials. Once the subject tract is selected, the extents of that feature can be used to pan the tract to the center of the map.

To control the properties of the data frame, create a data frame object. This object will also be used to list the layers within the data frame. As demonstrated in tutorial 3-1, with only one data frame, you will not need to worry about indexing or using a for loop designed to iterate through multiple data frames.

10. Add code to create a map document object, a data frame object, and a list object of all the layers included in the table of contents, as shown:

```
# Create a map document object for the currently open map document;
# a data frame list of the only data frame in this map document;
# a list object with a list of the layers in this map document.
thisMap = arcpy.mapping.MapDocument("CURRENT")
# Adding the index number at the end of the data frame object creation
# means that you are only working with the first data frame.
myDF = arcpy.mapping.ListDataFrames(thisMap)[0]
myLayers = arcpy.mapping.ListLayers(myDF)
```

Adding the index number at the end of the statement that creates the data frame list object means that you are dealing with only the first data frame (index number 0), which is okay because there is only one data frame. The result is that the myDF list object contains only one entry, and future use of the object will not require an index number. If you are working with a map document that has more than one data frame, you would not add [0] at the end of the creation statement, and the myDF object would have multiple entries. You may be able to use the activeDataFrame property of the data frame to place a single data frame in an object, but if you are unsure of the active data frame, you may not get the expected results. Then you would use a for loop to go through the list to find the particular data frame you wanted to work with.

With the layer list object created, you can go through that list with a for statement to find the layer named Parcels.

11. Add the code to scroll through the layer list object to find the Parcels layer. Optionally, you can add ArcPy messages to report back the progress while the script is running, as shown:

```
# Use a for loop to find the layer Parcels in the list
for lyr in myLayers:
    arcpy.AddWarning(lyr)
    if lyr.name == "Parcels":
        arcpy.AddWarning("Got it!")
```

It is interesting to see what the layer list object includes because the ArcMap table of contents contains both layers and group layers. Each group layer is shown by its name, and following good practice, all these group layers end with the word *group*, and all layers are prefaced by the group name. Note in the following graphic that the layer Lot Addresses appears in several groups. This naming convention is important to remember if you want a particular layer that may be repeated in other groups.

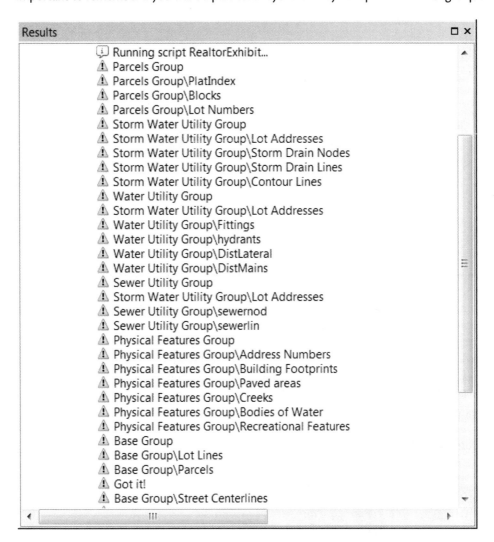

Next, do the feature selection.

12. **Add the code to select the feature using the tax account number and the field EKEY, as shown:**

```
# Do the selection process
arcpy.SelectLayerByAttribute_management(lyr,"NEW_SELECTION","EKEY = " + taxAccount)
```

13. Save your Python script, and run the tool to test it. You can enter anything you like for the first values, but for the tax account, enter the number 5561280.

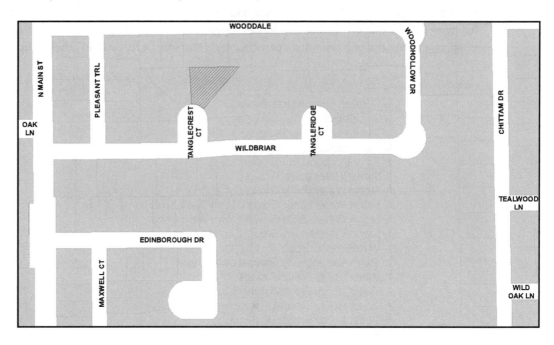

You see the parcel selected on the screen, as shown in the graphic. If not, go back and troubleshoot your code until this step works.

With the parcel selection operating successfully, the next step is to pan the map so that the selected feature is in the center of the map. There is a special method for layers to get the extent of the selected features and a variety of methods for data frames to pan or zoom to the selected feature's extent. Research these methods for the next step.

14. Add the code to pan the map to the selected features, as shown:

```
# Get the extent of the selected feature
lyrExtent = lyr.getSelectedExtent()
# Pan to the selected feature
myDF.panToExtent(lyrExtent)
# OPTION
# myDF.zoomToSelectedFeatures()
# Adjust the scale factor a bit
# myDF.scale = myDF.scale * 1.5
```

Note that the sample code includes an option to zoom the map instead of pan. If you want to try this option, change the extent property of the data frame to make zooming automatic. For this example, zooming in too closely to the parcel would not give a good overall impression of the utility systems, so the scale is fixed at 1:1200.

Make the script set the title and subtitle based on the user input.

Your turn

In tutorial 3-1, you learned how to access and change the text elements in a map document. Now add the code to set the map title and subtitle to the user's input. This code will involve making a list object of the map's elements and setting each object's text value according to its name value, as shown:

```
# Create a list of the map elements.
# These will be accessed by their names.
myElements = arcpy.mapping.ListLayoutElements(thisMap, "TEXT_ELEMENT")

# Scroll through the list of elements with a for statement.
# Use an if statement to test for each of the element names you
# are changing, and replace the text with the user input.
for element in myElements:
    if element.name == "Main Title":
        element.text = mainTitle
    elif element.name == "SubTitle":
        element.text = subTitle

# Refresh the map so that the changes can be seen.
    arcpy.RefreshActiveView()
```

The final step is to determine which boxes were checked in the user dialog box, turn layer groups on and off accordingly, and export the map exhibits.

The input dialog box allowed the user to select check boxes to determine which maps to draw. The choices were accepted by the script tool and sent to the variable mapList, which contains all the map names that were selected, separated by semicolons, as shown:

```
'Parcel Map';'Storm Water Utility Map';'Water Utility Map';'Sewer Utility Map'
```

To create the correct maps, you can interrogate this variable with an if statement, looking to see if it contains each individual map type. For example, the commands to make the water utility exhibit could be prefaced with an if statement to see if the words *Water Utility Map* were in this string.

15. Add an if statement to determine whether the check box for Parcel Map was selected, as shown:

```
# Find out if the check box for Parcel Map was selected
if "Parcel Map" in mapList:
    arcpy.AddWarning("Creating the Parcel Map...")
```

Your turn

Add if statements for each of the other map types to determine whether they were selected by the user, as shown:

```
# Find out if the check box for Water Utility Map was selected.
if "Water Utility Map" in mapList:
    arcpy.AddWarning("Creating the Water Map...")

# Find out if the check box for Sewer Utility Map was selected.
if "Sewer Utility Map" in mapList:
    arcpy.AddWarning("Creating the Sewer Map...")

# Find out if the check box for Storm Water Utility Map was selected.
if "Storm Water Utility Map" in mapList:
    arcpy.AddWarning("Creating the Storm Water Map...")
```

16. **Run the tool, select all the boxes, enter sample text for the titles, and use tax account number** 5561280. **Optionally, run the tool with only a couple of the check boxes selected, and note the results.**

With the process of choosing the correct map working, add the code to turn on or off the appropriate layers for each map. Refer to the description in the scenario for the correct layer combinations.

The process is simple. Use the layer list object that you have already created, and make a group layer visible or not visible by changing the visibility property. Remember to refresh the map and the table of contents.

17. **Add the code to set the visibility for all the layer groups in the case of the Parcel Map box being selected, as shown:**

```
# Find out if the check box for Parcel Map was selected.
if "Parcel Map" in mapList:
    arcpy.AddWarning("Creating the Parcel Map...")
    for lyr in myLayers:
        if lyr.name == "Parcels Group":
            lyr.visible = "True"
        if lyr.name == "Storm Water Utility Group":
            lyr.visible = "False"
        if lyr.name == "Water Utility Group":
            lyr.visible = "False"
        if lyr.name == "Sewer Utility Group":
            lyr.visible = "False"
        if lyr.name == "Physical Features Group":
            lyr.visible = "False"
        if lyr.name == "Base Group":
            lyr.visible = "True"
    arcpy.RefreshActiveView()
    arcpy.RefreshTOC
```

18. Add the line of code to export the map to a PDF document using the map type and the map title in the name, as shown in the graphic. Save the script when completed.

```
# Output completed map to a PDF document.
arcpy.AddWarning("Outputting the parcel map...")
arcpy.mapping.ExportToPDF(thisMap,r"C:\EsriPress\GISTPython\MyExercises\Parcel_Map_" + mainTitle)
```

19. Run the tool to test your progress. Select only the check box for Parcel Map. Enter a map title of Oak Forest Addition, a subtitle of Block A, Lot 14, and use tax account number 5561302. When the process is completed, open the PDF document in your MyExercises folder, and compare your file to the graphic shown:

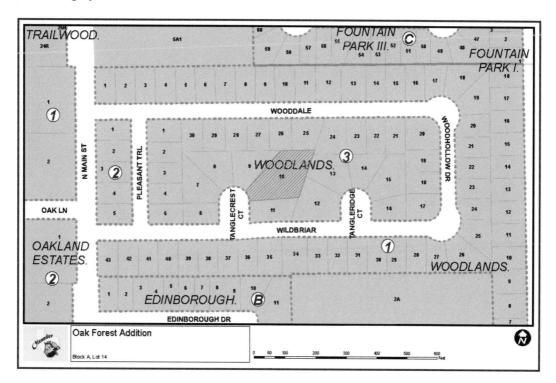

Your turn

Add the code to produce the other three map types. Control the visibility to turn the correct layer groups on, and output the map to a PDF document. Reference the description in the tutorial's scenario to see which layers to make visible for each map type.

20. Save the script, and close your IDE. Test the script tool using the information shown:

Map Types	Title	Subtitle	Tax Account
All	Royal Oaks Estates	Block 5, Lot 8	2571323
Water, Sewer	Woodcreek Addition	Block D, Lot 14	3599809
Sewer, Storm Water	Somerset Place	Block C, Lot 19	2792826
Sewer, Storm Water, Water	Alexander Addition	Block 10, Lot 1	18007

With this application completed, the task of making exhibit maps for real estate agents is now quite simple. It would be possible to allow the agents to fill out the form and create the maps themselves, or even use this application to develop a web-based entry form from which anyone could create an exhibit map and have it e-mailed back to them.

Here's the sample pseudo code for this project:

```
# Import the modules.
# Prompt the user for the maps to draw. This will be accepted as a list object.
# Prompt the user for a main title.
# Prompt the user for a subtitle.
# Prompt for a single tax account number.
# Create a map document object for the currently open map document;
# a data frame list of the only data frame in this map document;
# a list object with a list of the layers in this map document.
# Adding the index number at the end of the data frame object creation
# means that you are only working with the first data frame.
# Use a for loop to find the layer Parcels in the list.
# Do the selection process.
# Get the extent of the selected feature.
# Pan to the selected feature.
# Create a list of the map elements.
# These will be accessed by their names.
# Scroll through the list of elements with a for statement.
# Use an if statement to test for each of the element names you
# are changing, and replace the text with the user input.
# Refresh the map so that the changes can be seen.
# Find out if the check box for Parcel Map was selected.
# Find out if the check box for Water Utility Map was selected.
# Find out if the check box for Sewer Utility Map was selected.
# Find out if the check box for Storm Water Utility Map was selected.
# Output completed map to a PDF document.
```

Exercise 3-2

The city planner has seen the application for real estate agents and wants a similar application for property owner notification maps. He would like to be able to enter a list of tax account numbers (representing a single case) and have the application use these numbers to select the subject tracts. The application can automatically buffer the tracts 200 feet. Then he would like the application to zoom to that area and create a parcel map and a physical features map.

A sample map is provided, map document Exercise 3-2. The key differences in this application are as follows:

- The user will be able to enter more than one tax account number.
- The map will be zoomed rather than panned.
- There are two data frames in this map document.

Name all the elements in the map document, and prepare it for being automated. Then write the code, and create a script tool interface to accept account numbers and, optionally, a buffer distance (some cases may require more than a 200-foot buffer).

When the tool is finished running, you should have two maps for each case.

The first is a parcel map with the following layers:

- Lot_Boundaries
- PlatIndex
- Blocks
- Lot Numbers
- ZoningDistricts
- Street Names
- The new buffer layer that you create

The second is a physical features map with the following layers:

- Lot_Boundaries
- Address Numbers
- Building Footprints
- Paved areas
- Creeks
- Bodies of Water
- Recreational Features
- Street Names
- The new buffer layer that you create

Here are some planning cases that you can use for testing:

Tax Account Numbers	Map Title	Case Type	Property Description 1	Property Description 2
1784102, 1784110	RZ-04-2014	Rezone	SW Mills Addition	Block 2, Lots 4R and 5R
41433068	SP-12-2014	Site Plan	Fountain Center Addition	Lot 2
4647955, 653764, 653756	PL-09-2014	Plat	Cresthaven Addition	Block 1, Lots A, A1, and A2
2024888, 2024683	VR-11-2014	Variance	Oakland Estates	Block 2, Lots 2 and 19

Tutorial 3-2 review

This tutorial encompasses a variety of map manipulation techniques from tutorial 3-1 and many programming techniques from other chapters. These new techniques include the manipulation of the map's view of the data. In this case, the map was panned to the selected features, but it could also have been zoomed to the selected features and the scale adjusted.

The user input was also adapted to include as many data integrity techniques as possible. Presenting a quality interface and limiting the choices of the user increases the chances of a successful run.

Another new technique is handling the output of printable maps. These maps were exported to PDF documents, but other forms of presentation are available. It is important to make sure that the output file names are unique, which allows the user to create several sets of maps in one session and to know that the files are not being overwritten.

Study questions

1. What other data integrity techniques might be used to ensure that the tax ID number entered by the user is valid?

2. Why would zooming to selected features be better in some circumstances than panning to selected features?

3. Check ArcGIS for Desktop Help to determine the other types of map output formats.

Tutorial 3-3 Creating a map series

The techniques used for manipulating the map elements can be combined with geoprocessing tasks to create a series of custom maps. Feature lists and selections can be used to control the extent of the map.

Learning objectives

- Combining feature selection techniques and map element manipulations
- Cloning map elements
- Creating a map book

Preparation

Research the following topics in ArcGIS for Desktop Help:

- "GraphicElement (arcpy.mapping)": Clone
- "DataFrame (arcpy.mapping)": zoomToSelectedFeatures()
- "ExportToPDF (arcpy.mapping)"
- "PDFDocument (arcpy.mapping)"

Introduction

In tutorials 3-1 and 3-2, you learned how to manipulate the elements in a map document and how to control the properties of the layers in the table of contents. In this tutorial, you will combine selections and geoprocessing tasks with these other techniques to create a series of maps.

Making selections in a layer and iterating through the selected features is a common task. As each feature is accessed, the extent of the feature, or for that matter a set of features, can be used to pan or zoom the map. The panToExtent() method is used in tutorial 3-2 because the map was set to a fixed scale, but by using the zoomToSelectedFeatures() method, the scale of the map is changed to accommodate the selected features. One caveat is that the area covered by the selected set of features may vary greatly, so the map document must be able to accommodate many scales.

Although, in general, you are only able to work with elements already in the map document, and you can neither add nor delete elements, this limitation is not completely true because arcpy.mapping has the ability to clone graphic elements in your map to create copies that you can manipulate. Another common practice for using graphic elements is to include extra elements in the map document that are moved off the virtual page. If you need more elements than were originally used in the map document, make a list of all the elements in the map document, on or off the virtual page, and find one of the extras. Then move the element's location coordinates onto the virtual page. Elements not on the virtual page do not appear in prints or exported files.

Scenario

The Oleander Public Works Department must submit a water quality report to the State of Texas each year to maintain its ranking of "Superior Water Supply." Part of the process for developing the report is to take and test water samples at locations throughout the city. When the report is submitted, the department wants to include a series of small exhibit maps showing the location of each sampling station and a list of the property descriptions within 30 feet. It is supplying a point feature class with the sampling station locations, and it wants one map for each location.

Use the template map document provided, and iterate through each sampling station. Then select the properties within the specified distance, and generate the list of property descriptions. Make a table at the bottom of the map using graphic lines and text.

Each map should be exported to a PDF document, and when they are all completed, the maps can be combined to form a single map book.

Data

A map document is provided as a template, which includes the water utility system, the feature class containing the sampling stations, and a feature class containing the property information. The elements to use include the following:

- Water Utility—the data frame for the main map display.
- Title—the map title that you will change to reflect the station number.
- PropDesc—text that contains the property description of the selected properties. Clone this for each additional property.
- LeftVert and RightVert—the lines bordering the property description. Clone these for each additional property.
- TopHoriz and BotHoriz—the lines at the top and bottom of the property description. Clone the bottom line for each additional property.

In the sampling station feature class, the field named Status shows which stations are operational. Only these stations should be mapped. There is also a description that will appear on the map.

The layer Parcels has two fields to supply the legal description of the property: Prop_Des_1 and Prop_Des_2. Combine these two fields into a single text element for each parcel, and display this text on the exhibit map.

SCRIPTING TECHNIQUES

The new technique in this tutorial includes the functions to clone text and graphics. At least one of the graphic elements you wish to clone must be present in the map document. Then the item can be cloned using the clone method, and its location and size can be adjusted using the element's properties. It is a suggested practice to add a suffix value to the cloned items. This value can be any string you like, but _clone_ is customary. Each new feature gets this suffix along with a count value. For example, _clone_1, _clone_2, and _clone_3.

After you are finished with the map exhibit, use the delete method to delete the cloned items. These items will be easy to identify because of the suffix, so a ListLayoutElements results object with a wildcard can easily find them all. Deleting all the cloned elements resets the map for the next iteration.

Another technique is to establish a set of variables for each layout element. In the previous tutorials, a list was created and searched each time an element needed to be found and used, which can be resource intensive for a script that creates a large number of maps. Instead, go through the list of elements once, and establish variables for each element. Finding and modifying elements will be easier and faster after the variables are established.

The last new technique shown here is working with PDF documents. Besides exporting the map to a PDF document, you can use other methods with PDF documents. Rather than send a large number of PDF documents to another user, combine them into a single PDF document for added convenience. It is also a good way to create a single map book that can be easily shared and printed.

To accomplish this task, export the individual map pages as normal. Then create a new, empty PDF document to hold the final map book. Using a looping statement, append all the individual pages to the map book file, creating the desired single output file. As a matter of cleanup, delete the individual files so that only the map book remains.

Create a map series

1. **Start ArcMap, and open the map document Tutorial 3-3.**

Take a moment to examine the names of the graphic elements as described in the data section at the start of this tutorial. You can use this map as a reference, but this script is a stand-alone script and has no user parameters. When it runs, the script creates a PDF document of each map and, at the end, combines all the maps into a single PDF document.

2. **Write the pseudo code for this script. Yours will be more detailed, but follow this basic outline:**

- Iterate through the features in the SamplingStations layer, making sure to use only the active station.
- Use the current sampling station to select all the property within 30 feet.
- Zoom the map document to show the currently selected features.
- Change the map title to reflect the sampling station.
- Build a chart of the property descriptions, and separate the values with lines.
- Create an exhibit map for each sampling station, and when completed, merge all the maps into a single PDF document.

When you have your pseudo code ready, continue with the tutorial.

3. **Start your IDE, and create a new script named** SampleStationsMapBook.py **in your MyExercises folder.**

4. **Add the title information and author, and import the ArcPy module, as shown:**

```
#-------------------------------------------------------------------------
# Name:         Create Sampling Stations Exhibit Map
#
# Purpose:      Individual exhibit maps will be created for each sampling station
#               in Oleander. These will show the location and a list of the property
#               within 200 feet. When completed, all the individual maps will be
#               combined into a single map book.
#
# Author:       David W. Allen, GISP
#
# Created:      10/25/2013
# Copyright:    (c) David 2013
#
#-------------------------------------------------------------------------


# Import the modules
import arcpy
```

5. **Create the reference to the map document with an object named** thisMap**, as shown in the graphic. Note that the full path is given so that this script can be run as a stand-alone script.**

```
# Create a map document object for the map document
thisMap = arcpy.mapping.MapDocument(r"C:\EsriPress\GISTPython\Maps\Tutorial 3-3.mxd")
```

6. Create a data frame object to reference the Water Utility data frame. This method shows a way to find the correct data frame with a single index, even if it is not the first data frame (index[0]), as shown:

```
# Create a data frame list object of the Water Utility data frame
# Adding the index number at the end of the data frame object creation
# means that you are working with only the desired data frame
myDF = arcpy.mapping.ListDataFrames(thisMap)[0]
if myDF.name <> "Water Utility":
    myDF = arcpy.mapping.ListDataFrames(thisMap)[1]
```

7. Use the ListLayoutElements method to create a list object of all the graphic elements in the map. Then use a for statement to loop through the elements to assign each element to a variable, as shown:

```
# Create a list object of the map elements
myMapElements = arcpy.mapping.ListLayoutElements(thisMap)

# Use a loop to assign the elements you want to control to variables
for element in myMapElements:
    if element.name == "Title":
        mapTitle = element
    elif element.name == "PropDesc":
        mapDesc = element
    elif element.name == "LeftVert":
        mapLeftVert = element
    elif element.name == "RightVert":
        mapRightVert = element
    elif element.name == "TopHoriz":
        mapTopHoriz = element
    elif element.name == "BotHoriz":
        mapBotHoriz = element
print mapTitle.text
print mapDesc.text
```

This technique performs the loop through the element list only once, saving time and code as the elements are referenced throughout the script. Note the inclusion of print statements to track the progress of the script.

8. Create a list object of the layers in the table of contents. Find the Parcels and SamplingStations layers, and assign them to variables for future use. Create a feature layer of the sampling stations with a wildcard to include only the operational stations, as shown:

```
# Get a list of all the layers in the table of contents
myLayers = arcpy.mapping.ListLayers(myDF)

# Use a for loop to find the layers 'Parcels'
# and 'SamplingStations' in the list
# Assign them to new variables for easier reference

for lyr in myLayers:
    if lyr.name == "Parcels":
        print "Found the parcels!"
        layerParcels = lyr
    elif lyr.name == "SamplingStations":
        print "Found the sampling stations!"
        layerSStations = lyr

# Create a feature layer to hold only the active sampling stations
arcpy.MakeFeatureLayer_management(layerSStations,"ActiveStations_lyr",'"Status"=\'Operational\'')
```

The feature layer of the sampling stations is used to access each station for the maps. Note that the Parcels layer is not being put into a feature layer. Although you can do selections in a feature layer, you cannot use them to manipulate the map extent or scale. Because feature layers exist in virtual memory, their connections to the map document and data frame are removed. Changing the extent or scale of a feature layer does not affect the map document. You may, however, get the extent of a feature layer and use it to set the data frame extent, which would have the effect of zooming to the feature layer.

9. **Add the code to create a new folder in your MyExercises folder named** WaterExhibitMaps **to hold the output PDF documents, as shown:**

```
# Create a new folder to hold the output PDF documents
arcpy.CreateFolder_management(r"C:\EsriPress\GISTPython\MyExercises", "WaterExhibitMaps")
```

Each completed map document is exported to this folder as a PDF document. At the end of the script, access this folder to combine all the files into one PDF document.

The script so far includes the code to set up all the data and workspaces for the geoprocessing and map creation tasks to be done. Next, set up a cursor to step through the features in the sampling stations feature layer. Feature selections and map document manipulations will be made for each feature, producing an individual map for each station.

10. **Set up a search cursor for the sampling stations feature layer. Include the field** Desc **with the cursor to be used as the map title. Write a for statement to iterate through each feature, as shown:**

```
# Create a cursor object for the sampling stations
stationCursor = arcpy.da.SearchCursor("ActiveStations_lyr","Desc")

# Iterate through the features of the sampling stations
for feature in stationCursor:
```

The field Desc will be used to name each output map. The cursor object will contain a single attribute value for each feature, which means that index number 0 will hold the value from that field. If more fields were included in the cursor initialization, the fields would be accessed in the order in which they were called in the cursor command, with index numbers starting at one.

11. **Set a variable named** featureDesc **to contain the attribute value for the current feature, as shown:**

```
# Retrieve the station name from the current feature
featureDesc = feature[0]
```

The current feature in the cursor is referenced but not selected. To perform the selection with the 30-foot buffer, select the currently referenced feature, which can be done using the query "Desc" = 'featureDesc.' Selecting this single item makes it easy to select by location.

12. Add a select by attribute function to select the currently referenced feature. Use a select by location function to select all the parcels within 30 feet, as shown:

```
# Select the current feature in the Parcels layer
# SelectLayerByAttribute_management (in_layer_or_view,
#           {selection_type}, {where_clause})
arcpy.SelectLayerByAttribute_management("ActiveStations_lyr","NEW_SELECTION",\
'"Desc"=\'' + featureDesc + "'")

# Select the parcels within 30 feet of this feature
# SelectLayerByLocation_management (in_layer, {overlap_type},
#           {select_features}, {search_distance}, {selection_type})
arcpy.SelectLayerByLocation_management(layerParcels,"WITHIN_A_DISTANCE",\
"ActiveStations_lyr","30")
```

With the parcel feature selected, you can alter the extent of the map to show the current sampling station and the parcels around it. Accessing the extent property of the selected features and passing this along to the data frame would be fine, but the zoomToSelectedFeatures() method accomplishes this task with one line of code. Because the sampling station selection was done on a feature layer, it will have no effect on the map extent. The parcel selection, which was done on the layer in the table of contents, will control the zoom.

Once the zoom is completed, check the scale to make sure that it is not zooming in too closely. A scale of 1:1800 should be the closest zoom, and if the scale factor is larger than that, it is a good idea to increase the scale by about 10 percent. This change keeps the items at the edge of the selected set from touching the map border. Note that the arcpy.RefreshActiveView or arcpy.RefreshTOC functions are not necessary here. These functions refresh the screen and have no effect on a stand-alone script.

13. Add the code to zoom to the selected features. Then test to see if the scale factor is below 1800. If so, set the scale to 1800, and if the scale is larger, add a 10 percent margin to the scale, as shown:

```
# Zoom the map to the selected features
myDF.zoomToSelectedFeatures()
# Adjust scale to at least 1:1800
if myDF.scale < 1800:
    myDF.scale = 1800
else:
    myDF.scale = myDF.scale * 1.1
```

The map area is now set to show the desired features at an appropriate scale. The next phase is to set the map title and to build a chart showing the descriptions of the selected parcels. Refer back to the variable names assigned to the different graphic items, noted when you first opened the map document, to get the names of the objects you will change. The graphic storing the map title was set to the variable mapTitle, so to change its text, access the mapTitle.text property.

14. Set the map title to the current station description, stored in the featureDesc variable, as shown:

```
# Set the text in the map
# Change the title of the map to this name
mapTitle.text = featureDesc
```

The chart consists of the property description and the three lines abutting the text. If you examine the map document, you will see these graphic items below the virtual page. In this location, these items will not appear on any printed or exported maps. Clone these items, and then move the clones into place to represent each property that is selected. Earlier in the script, the elements were given the variable names mapLeftVert, mapRightVert, and mapDesc. Each element has a property for the X and Y page coordinates named elementPositionX and elementPositionY.

To create the chart, set up variables to hold the page coordinates for the first set of lines and the first chart text derived from the first parcel. Clone the elements, and move them to these positions. For each additional parcel, move the page coordinates down 0.25 inches, and place a new set of cloned items in that position.

15. Set up variables to hold the coordinates for the first set of lines and text using the coordinate values shown:

```
# Set up the locations for the lines and text used in the chart
# For each text value, the existing items will be cloned and used
# in the map.

# Set up the coordinates for the lines and text of the chart
botLineX = 0.32
botLineY = 2.00
leftLineX = 0.32
leftLineY = 2.25
rightLineX = 3.41
rightLineY = 2.25
chartTitleX = 0.38
chartTitleY = 2.21
```

Many parcels may be selected for each map. To get all the proper descriptions into the charts, set up a cursor to scroll through the selected properties. This parcel cursor is nested inside the sampling station cursor. For each sampling station, the parcel cursor will return multiple parcels. When the end of the parcel cursor is reached, the sampling station cursor will move to the next feature, and the process will start again.

16. Set up a search cursor for the parcels. Have the cursor return the fields Prop_Des_1 and Prop_Des_2, which will be used in the chart. Then set up the for statement to step through the selected parcels, as shown:

```
# Get the values of the property description and build chart
fieldList= ["Prop_Des_1","Prop_Des_2"]
parcelCursor = arcpy.da.SearchCursor(layerParcels,fieldList)
for label in parcelCursor:
```

The text element containing the property description can be cloned into an object named textClone. The suffix _clone makes it easier to select and delete the cloned items before the next map is created. Once the item is cloned, it is given the concatenated value of the two property descriptions. Then it is moved into position using the coordinates shown in the preceding graphic.

17. Add the code to clone the element PropDesc, set it to the concatenation of the two property descriptions from the selected parcels, and move it into position, as shown:

```
# Clone the chart title and set the text
textClone = mapDesc.clone("_clone")
textClone.text = str(label[0]) + ", " + str(label[1])
# Move the new text into place
textClone.elementPositionX = chartTitleX
textClone.elementPositionY = chartTitleY
```

18. Write the code to perform a clone operation on the three lines used to make the chart, and move them into place, as shown:

```
# Clone the lines and move them into place
leftLine = mapLeftVert.clone("_clone")
rightLine = mapRightVert.clone("_clone")
botLine = mapBotHoriz.clone("_clone")
leftLine.elementPositionX = leftLineX
leftLine.elementPositionY = leftLineY
rightLine.elementPositionX = rightLineX
rightLine.elementPositionY = rightLineY
botLine.elementPositionX = botLineX
botLine.elementPositionY = botLineY
```

The first line of the chart is finished. If there are more parcel descriptions to add, the parcelCursor repeats, and another feature is selected. In order for this line in the chart to appear in the right location, move everything down 0.25 inches, which involves subtracting that value from the y-values of the lines and text. The x-values remain unchanged. This adjustment should be made before the cursor repeats.

Once all the items in the cursor have been processed, delete the cursor object, which is done by going back to the same indent level as the for statement and issuing the del command.

19. Add the code to reduce all the y-coordinates by 0.25 inches, and delete the cursor, as shown:

```
        # Move the Y coordinate values down 0.25 inches
        botLineY = botLineY - 0.25
        leftLineY = leftLineY - 0.25
        rightLineY = rightLineY - 0.25
        chartTitleY = chartTitleY - 0.25
    del parcelCursor
```

When the parcel cursor is finished running, the map for the first sampling station will be complete. Export the map to a PDF document before repeating the process for the next station.

20. Export the map to a PDF document into the folder you created in step 9, as shown:

```
# Export the map to a PDF document
# ExportToPDF (map_document, out_pdf, {data_frame}, {df_export_width},
#    {df_export_height}, {resolution}, {image_quality},
#    {colorspace}, {compress_vectors}, {image_compression}, {picture_symbol},
#    {convert_markers}, {embed_fonts},
#    {layers_attributes}, {georef_info}, {jpeg_compression_quality})
arcpy.mapping.ExportToPDF(thisMap, r"C:\EsriPress\GISTPython\MyExercises\WaterExhibitMaps\Project_" \
    + mapTitle.text + ".pdf", resolution=100,image_quality="NORMAL")
```

In this example, the file resolution and image quality are reduced to save file space. You can research the other settings in ArcGIS for Desktop Help.

When the map has been exported, remove all the cloned items from the map in preparation for the next map. Because the clones all have the suffix _clone, they can be easily identified using a wildcard with the ListLayoutElements() function. Use a for statement to iterate through the list and delete the elements.

21. Create a list object with the cloned elements, and use a for statement to step through and delete the elements, as shown:

```
# Delete the cloned items and reset for the next map
cloneGraphics = arcpy.mapping.ListLayoutElements(thisMap,wildcard="*clone*")
for graphic in cloneGraphics:
    graphic.delete()
```

This is the end of the sampling stations cursor. A complete map has been created, and the map document has been reset for the next map. The cleanup at the end of the cursor's process is to delete the cursor and the map document you have been accessing. Note that these return to the same indent level as the for statement you used to step through the sampling stations cursor.

22. **Add the code to close the cursor and the map document, as shown:**

```
# Delete the cursor
del stationCursor

# Delete the map document object
del thisMap
```

Take all the PDF documents created by this script and combine them into a single PDF document. The "Getting Started with arcpy.mapping" tutorial in ArcGIS for Desktop Help has a section demonstrating this process, and the code shown in the graphic for step 23 is derived from that example.

23. **Combine all the PDF documents in the WaterExhibitMaps folder into a single PDF document. This process involves creating the file, setting the workspace, creating a list of all the PDF documents in the destination folder, iterating through that list, and appending each file to the final document, as shown:**

```
# Combine all the PDF documents into a single map book
finalPDF_filename = r"C:\EsriPress\GISTPython\MyExercises\WaterExhibitMaps\StationsMapBook.pdf"
finalPDF = arcpy.mapping.PDFDocumentCreate(finalPDF_filename)

# Create a list of all PDF documents in the new folder
arcpy.env.workspace = r"C:\EsriPress\GISTPython\MyExercises\WaterExhibitMaps"

# Copy each file with a .pdf extension to a dBASE file
# List automatically looks in the defined workspace
pdfPath = r"C:\EsriPress\GISTPython\MyExercises\WaterExhibitMaps\\"
for file in arcpy.ListFiles("*.pdf"):
    print file
    finalPDF.appendPages(pdfPath + file)

# Commit changes and delete variable reference
finalPDF.saveAndClose()
del finalPDF
```

24. Save the SampleStationsMapBook.py script. Close the map document, and run the script. When the script is finished, look in the destination folder for the output PDF document.

Here's the pseudo code for this script:

```
# Import the modules
# Create a map document object for the map document
# Create a data frame list object of the Water Utility data frame
# Adding the index number at the end of the data frame object creation
# means that you are working with only the desired data frame
# Get a list of all the layers in the table of contents
# Create a list object of the map elements
# Use a loop to assign the elements you want to control to variables
# Use a for loop to find the layers 'Parcels'
# and 'SamplingStations' in the list
# Assign them to new variables for easier reference
# Create a feature layer to hold only the active sampling stations
# Create a new folder to hold the output PDF documents
# Create a cursor object for the sampling stations
# Iterate through the features of the sampling stations
# Retrieve the station name from the current feature
# Select the current feature in the Parcels layer
# SelectLayerByAttribute_management (in_layer_or_view, {selection_type}, {where_clause})
# Select the parcels within 30 feet of this feature
# SelectLayerByLocation_management (in_layer, {overlap_type}, {select_features}, \
#           {search_distance}, {selection_type})
#
# Zoom the map to the selected features
# Adjust scale to at least 1:1800
# Set the text in the map
# Change the title of the map to this name
# Set up the locations for the lines and text used in the chart
# For each text value, the existing items will be cloned and used
# in the map.
# Set up the coordinates for the lines and text of the chart
# Get the values of the property description and build chart
# Clone the chart title and set the text
# Move the new text into place
# Clone the lines and move them into place
# Move the Y coordinate values down 0.25 inches
# Export the map to a PDF document
# ExportToPDF (map_document, out_pdf, {data_frame}, {df_export_width}, {df_export_height}, \
#           {resolution}, {image_quality},
#           {colorspace}, {compress_vectors}, {image_compression}, {picture_symbol}, \
#           {convert_markers}, {embed_fonts},
#           {layers_attributes}, {georef_info}, {jpeg_compression_quality}))
# Delete the cloned items and reset for the next map
# Delete the cursor
# Delete the map document object
# Combine all the PDF documents into a single map book
# Create a list of all PDF documents in the new folder
# Copy each file with a .pdf extension to a dBASE file
# List automatically looks in the defined workspace
# Commit changes and delete variable reference
```

This script is rather complex, but when it works correctly, you can automatically make dozens of maps.

Exercise 3-3

Search ArcGIS for Desktop Help for the topic "Building map books with ArcGIS." Work through these examples, and research the ways the Data Driven Pages commands can be used in a Python script.

Tutorial 3-3 review

The mapping module in ArcPy is useful in creating map books, even if the page layouts are different from one map to another. The use of a cloning technique allows you to place extra graphic elements in a map layout and use these elements to build such things as charts, tables, and graphs. A graphic item can be cloned and moved into position as many times as needed.

Another interesting technique demonstrated in this tutorial is the process of creating a list of graphic elements and assigning the elements you wish to work with to a variable. This means that your cursor has to iterate through the list of elements only once, which can save a lot of time if you are doing a lot of work with elements.

You also learned the variety of tasks that can be performed involving PDF documents. Several map pages, title sheets, or even indexed values, such as street names, can be placed into separate PDF documents, and then these pages can be grouped into a single PDF document for easy printing and sharing.

Study questions

1. Would it be possible to nest more than one cursor (iteration) within another? What are the consequences of trying to do multiple iterations?
2. What other functions can be used to handle PDF documents in arcpy.mapping? Give examples.
3. What other uses might you see for the cloning technique?

Chapter 4

Python toolboxes

Introduction

The scripts and script tools you have written so far can be powerful and useful, but they are not easily portable or shared. Often, you need to either share multiple files or know that the recipient is able to open and run Python scripts in an IDE. These options are not user-friendly and would be prohibitive if you were sharing your custom tools with hundreds of users.

Two new features were added in ArcGIS 10.1 to help you write better applications and share them easily with others—Python toolboxes and Python add-ins. This chapter covers Python toolboxes, showing you the benefits of this new feature and how to create and share these toolboxes. Python add-ins are covered in chapter five.

Tutorial 4-1 Creating a Python toolbox

Python toolboxes were introduced in ArcGIS 10.1 with the express purpose of making your script tools easier to program and to share.

Learning objectives

- Understanding Python toolbox components
- Creating and sharing toolboxes
- Defining multiple tools

Preparation

Research the following topics in ArcGIS for Desktop Help:

- "What is a Python toolbox?"
- "Comparing custom and Python toolboxes"
- "Creating a new Python toolbox"
- "Editing a Python toolbox"

Introduction

The Python toolbox has two main advantages over a custom toolbox when it comes to building your own tools. The first advantage is that tools in Python toolboxes are easier to develop because all the code is stored in a single file. A script tool requires a Python script (.py) file to store executable code and a custom toolbox (.tbx) file to store the tool's interface. The user must manually coordinate the code with the toolbox properties for inputs, filters, value lists, and output. A Python toolbox contains the Python code for all the tools and interfaces in a single Python toolbox (.pyt) file, making the scripting simpler to write and share.

The Python toolbox code structure contains different classes. Each class handles a different tool, so your toolbox can contain multiple tools. Within the classes are modules that contain an aspect of the code that may deal with the start-up environment, the input and output parameters, and the business logic for the tools. There are also separate modules to handle validation messages and to check the license and extension availability when any of the tools in the toolbox are used. With all these processes handled in a single file, these applications are easier to code and troubleshoot.

The second advantage is portability. If you are sharing an application that is written as a standard script tool, multiple files must be moved to transfer the application to another machine. If there are issues in the code after the move, both files would need to be scrutinized to find the errors. However, a Python toolbox application can be moved as a single file, and it presents a single set of code for troubleshooting.

There is, however, a downside to the Python toolbox. It cannot contain regular script tools or models—these remain the exclusive domain of custom toolboxes.

Scenario

The application you wrote in tutorial 2-4 is working fine, and you would like to share it with all the members of the library. If you recall, this application took a previously selected site and determined how far out you would have to go to find 200 registered library patrons. Sharing the application will involve putting it on 15 to 25 different machines, so you want to make the tool as portable as possible. To accomplish this task, take the code written for tutorial 2-4, and migrate it to a Python toolbox.

Data

A completed copy of the code is in text file Tutorial 4-1 in the Data folder. This code will access the same data as tutorial 2-4.

SCRIPTING TECHNIQUES

The code for the Python toolbox is separated into several parts, each of which controls a different aspect of the tools you add to the toolbox. The first part is the Toolbox class. The code for the Toolbox class looks like this:

```
import arcpy


class Toolbox(object):
    def __init__(self):
        """Define the toolbox (the name of the toolbox is the name of the .pyt file)."""
        self.label = "Toolbox"
        self.alias = ""

        # List of tool classes associated with this toolbox
        self.tools = [Tool]
```

Import the ArcPy module.

The class Toolbox(object) contains information about the toolbox and the names of all the tools the toolbox will contain. The class name Toolbox is protected and should not be changed.

Within this class, define a toolbox label and a toolbox alias—the same things you would set for a regular toolbox. Make a list of all the class names for the tools this toolbox will contain. For instance, a toolbox for public works that had tool classes named StreetArea, WorkNotify, and StormFlow would have the code shown:

```
import arcpy


class Toolbox(object):
    def __init__(self):
        """Define the toolbox (the name of the toolbox is the name of the .pyt file)."""
        self.label = "Public Works"
        self.alias = "pworks"

        # List of tool classes associated with this toolbox
        self.tools = [StreetArea,WorkNotify,StormFlow]
```

Note the list object self.tools that contains a list of all the tool class names within the toolbox.

Your public works Python toolbox would display the tool labels for each class, such as Storm Drain Flow Analysis, Street Area Calculation, and Work Notification, and look like this:

For each tool listed, there is a class containing a set of code to define and control it. The three main components of this class are the initialize (_init_) module, the parameter module, and the code module.

The _init_ module allows you to set a tool label and description and control the availability of background processing. The code for the StreetArea class in the public works example looks like this:

```
class StreetArea(object):
    def __init__(self):
        """Define the tool (tool name is the name of the class)."""
        self.label = "Street Area Calculation"
        self.description = "The user selects a street and the area of pavement is calculated." + \
                            "The user can specify the measurements in square feet or square yards"
        self.canRunInBackground = False
```

Note that the class defines a label and a description that will be displayed in the toolbox.

Each tool in a Python toolbox is placed either at the root level of the toolbox or into a subcategory called a *toolset* and is always displayed in alphabetical order. Toolsets can be used to help organize tools within a toolbox, making them easier for the user to find. Toolsets are defined in the _init_ module of a tool's class by setting the self.category parameter, as shown in the following graphic. You can have as many tools as you like in a toolset by adding the same category statement to the tool's _init_ module.

```
class StreetArea(object):
    def __init__(self):
        """Define the tool (tool name is the name of the class)."""
        self.label = "Street Area Calculation"
        self.description = "The user selects a street and the area of pavement is calculated." + \
                            "The user can specify the measurements in square feet or square yards"
        self.canRunInBackground = False
        self.category = "Street Tools"
```

In the following graphic, toolsets were created for Storm Water and Street Tools. Each toolset can contain multiple tools. The tool Work Notification was left outside of a toolset and appears at the root level of the main toolbox, as shown:

The next component of the tool's class is the parameters module. In this module, all the input and output parameters are set in much the same way as creating a script tool. For each parameter, many properties can be set, including the input name and label, data type, parameter type, and direction. A complete list of the parameter properties and methods can be found in ArcGIS for Desktop Help by searching for "Defining parameters in a Python toolbox" or "Defining parameter data types in a Python toolbox." As an example, the code to accept two parameters for the public works tool would look like this:

```
def getParameterInfo(self):
    """Define parameter definitions"""
    # First parameter
    param0 = arcpy.Parameter(
        displayName="Input Features",
        name="in_features",
        datatype="DEFeatureClass",
        parameterType="Required",
        direction="Input")

    # Second parameter
    param1 = arcpy.Parameter(
        displayName="Map Title",
        name="mapitle",
        datatype="GPString",
        parameterType="Optional",
        direction="Input")
    # Set a default value
    param1.value = "Public Works Map"

    params = [param0, param1]
    return params
```

Index numbers are used with the parameters the same as in a script tool, and you can set many of the same properties. The list of valid input data types is basically the same as that used for creating a script tool, but the keywords are different for

Python toolboxes. In this example, the feature class data type is DEFeatureClass, and the string data type is GPString. Refer to the ArcGIS for Desktop Help topic "Defining parameter data types in a Python toolbox" for the complete list. The getParameterInfo method returns a list object you define containing the tool's parameters, which can be referenced in the code section of the tool's class.

When the tool runs, each parameter listed appears in the input dialog box and allows user interaction, as shown, which basically replaces the arcpy.GetParameterAsText() command used to get input from the user in a script tool.

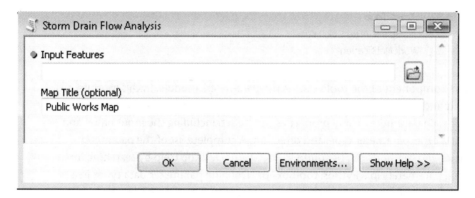

One of the most useful properties that can be set in the parameters module is a filter. In the same way that you controlled the input in a script tool, a filter can be set to control the input in a Python toolbox tool. A simple filter might be to set up a value list. The code for setting a value list filter identifies the filter type and defines the list of values. The code shown in the following graphic is setting the filter type to ValueList and then creating a list object with the valid values. A comment in this code lists all the valid filter types, which must be entered as shown:

```
class WorkNotify(object):

    def getParameterInfo(self):
        """Define parameter definitions"""
        param0 = arcpy.Parameter(
            displayName="Material",
            name="pipeMat",
            datatype="GPString",
            parameterType="Required",
            direction="Input")
        # Set the filter type
        # Valid filter types are 'ValueList', 'Range', 'FeatureClass', 'File', 'Field', and 'Workspace'
        param0.filter.type = 'ValueList'
        # Define a list object with the choices
        param0.filter.list = ['PVC', 'Clay', 'Steel', 'AC Conc']
        params = [param0]
```

Another interesting filter might be to set a field name selection that is dependent on a selected feature class. For instance, the user would select a feature class in a tool's first parameter, and the second parameter would present a value list of field names from that feature class. In the example code shown in the following graphic, the first parameter has a data type set to feature class and a filter set to Polygon.

This filter allows the selection of only polygon-type feature classes. The second parameter has a data type of Field, a filter of Text, and a dependency set to the value of the first parameter, which would allow only string fields from the selected feature class to appear in the value list.

```python
def getParameterInfo(self):
    """Define parameter definitions"""
    param0 = arcpy.Parameter(
        displayName="Input Features",
        name="inFeat",
        datatype="DEFeatureClass",
        parameterType="Required",
        direction="Input")
    # Limit the choice of feature class type to polygons
    param0.filter.list = ['Polygon']

    param1 = arcpy.Parameter(
        displayName="Destination Field",
        name="destField",
        datatype="Field",
        parameterType="Required",
        direction="Input")
    # List a specific field type
    param1.filter.list = ['Text']
    # Define a dependency
    param1.parameterDependencies = [param0.name]
    # Only string fields from the selected feature class will be shown

    params = [param0, param1]
    return params
```

Many more filter settings are available for establishing data integrity rules. These settings are found in the ArcGIS for Desktop Help topic "Defining parameters in a Python toolbox." The Help topic lists valid filter types, keywords for data types and field types, and sample code for their use. Remember that the more that you use data integrity rules to control data entry, the fewer errors users will have when using your tools.

The final component of a tool's class is the source code, or execute, module. The execute module contains the code that the tool will use to carry out whatever processes your application requires. In the example shown in the following graphic, two parameters were defined in the parameters module and were then brought into variables in the execute module by referencing their index numbers. The variables can then be used in the code.

```python
def execute(self, parameters, messages):
    """The source code of the tool."""
    inputFC = parameters[0].valueAsText
    destField = parameters[1].valueAsText
    # Your executable code goes here. #
    return
```

Notice that the name references for the parameters have changed. When these parameters are defined in the getParameterInfo method, you might use param0 and param1. These parameters are then returned to the script as a list object named parameters[], and they keep the same order and thus the same index numbers. Note that the definition line of the execute module calls the parameters list object, so, for example, param0 becomes parameters[0], and param1 becomes parameters[1].

Each Python toolbox includes a class to define the characteristics of the toolbox itself and a class for each tool it contains. Each tool class, at a minimum, has an _init_ module to define the tool and an execute module to contain the source code. It is also common to have a parameters module to define input and output parameters for the tool, if it has any. Other modules can be used to define validation rules and messages.

EDITING PYTHON TOOLBOXES

Most IDEs do not recognize a Python toolbox .pyt file as a script file, and therefore do not perform code completion or syntax-checking functions. Refer to appendix A for information on how to configure your particular IDE to handle .pyt files.

Also remember to set the Geoprocessing Options in ArcMap to whichever IDE you wish to use. If no preference is set, your operating system may open a text editor that is not suitable for editing Python code.

Create a Python toolbox

1. **Open the map document Tutorial 4-1, and pin the Catalog window to the desktop interface.**

   Shown are the parcels for Oleander and the suggested bookmobile sites that the head librarian suggested.

2. **In a text browser, open the text file Tutorial 4-1 in the Data folder.**

   This text file, shown in the graphic, is a completed version of the script you created in tutorial 2-4 to perform the library patron analysis.

```python
# Import the modules.
import arcpy

# Set up the environment.
arcpy.env.workspace = r"C:\EsriPress\GISTPython\Data\City of Oleander.gdb\\"
arcpy.env.overwriteOutput = True

# Set up cursor for the bookmobile sites.
arcpy.MakeFeatureLayer_management("Parcels","Parcels_Lyr",'"DU"= 1')
arcpy.MakeFeatureLayer_management("BookmobileLocations","Locations_lyr")

siteCursor = arcpy.da.SearchCursor("Locations_lyr","Marker")
for row in siteCursor:
    siteName = row[0]

    # Select parcels within 150 feet and store as a new selection.
    arcpy.Select_analysis("Locations_lyr",\
    r"C:\EsriPress\GISTPython\MyExercises\Scratch\TemporaryStorage.gdb\SiteTemp",\
    '"Marker" = \'' + siteName + "\'")

    arcpy.SelectLayerByLocation_management("Parcels_lyr","WITHIN_A_DISTANCE",\
    r"C:\EsriPress\GISTPython\MyExercises\Scratch\TemporaryStorage.gdb\SiteTemp","150","NEW_SELECTION")

    # Start a while statement until number of dwelling units exceeds 200.
    parcelCount = int(arcpy.GetCount_management("Parcels_lyr").getOutput(0))
    print parcelCount

    myCount = 1
    while parcelCount < 200:
        # All statements at this indent level are part of the while loop.

        # Add to the selected set all property within 150 feet and redo count.
        # SelectLayerByLocation_management (in_layer, overlap_type, select_features, search_distance, selection_type)
        arcpy.SelectLayerByLocation_management("Parcels_lyr","WITHIN_A_DISTANCE","Parcels_lyr","150","ADD_TO_SELECTION")

        parcelCount = int(arcpy.GetCount_management("Parcels_lyr").getOutput(0))
        print parcelCount
        if myCount == 8:
            parcelCount = 200
        else:
            myCount = myCount + 1

        # Exit the while statement when count exceeds 200.

    # Export the selected features to a new feature class using the Marker name.
    arcpy.CopyFeatures_management("Parcels_lyr", r"C:\EsriPress\GISTPython\MyExercises\MyAnswers.gdb\\" + siteName.replace(" ","_"))
    print siteName + " Output OK!"
    # Move to the next site and repeat.
```

To create a Python toolbox application from this script, create a new Python toolbox, and configure it to have the source code from the provided file. Creating the new toolbox is simple because it requires only a name.

3.   **In the Catalog window, right-click your MyExercises folder and click New > Python Toolbox, as shown. Name the toolbox** Proximity Tools.pyt.

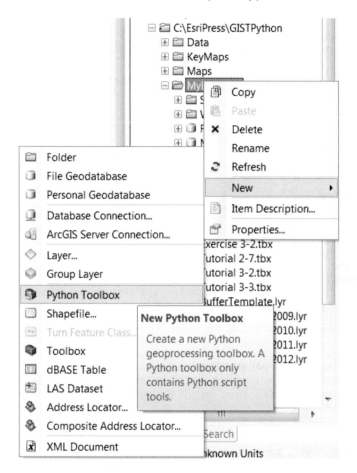

Next, edit the code for the Python toolbox. In a standard custom toolbox, right-click the script tool and click Edit because you will be editing each script separately. In a Python toolbox, right-click the toolbox and click Edit because all the code for all the tools this toolbox contains is in the single .pyt file.

4. Right-click Proximity Tools.pyt and click Edit. The code is displayed as shown:

```python
import arcpy


class Toolbox(object):
    def __init__(self):
        """Define the toolbox (the name of the toolbox is the name of the
        .pyt file)."""
        self.label = "Toolbox"
        self.alias = ""

        # List of tool classes associated with this toolbox
        self.tools = [Tool]


class Tool(object):
    def __init__(self):
        """Define the tool (tool name is the name of the class)."""
        self.label = "Tool"
        self.description = ""
        self.canRunInBackground = False

    def getParameterInfo(self):
        """Define parameter definitions"""
        params = None
        return params

    def isLicensed(self):
        """Set whether tool is licensed to execute."""
        return True

    def updateParameters(self, parameters):
        """Modify the values and properties of parameters before internal
        validation is performed.  This method is called whenever a parameter
        has been changed."""
        return

    def updateMessages(self, parameters):
        """Modify the messages created by internal validation for each tool
        parameter.  This method is called after internal validation."""
        return

    def execute(self, parameters, messages):
        """The source code of the tool."""
        return
```

The default template that was created with the Python toolbox contains the classes and modules you need for the tools. Configure the toolbox name, description, and the list of tools it will contain. This configuration is done in the Toolbox class.

5. In the Toolbox class, set the label to Proximity Tools and the alias to prox. Then enter the tool name PatronSelect in the self.tools list object, replacing the generic "tool," as shown:

```python
class Toolbox(object):
    def __init__(self):
        """Define the toolbox (the name of the toolbox is the name of the .pyt file)."""
        self.label = "Proximity Tools"
        self.alias = "prox"

        # List of tool classes associated with this toolbox
        self.tools = [PatronSelect]
```

Now you can change the Tool class to accept the parameters for your library tool.

6. Change the name of class Tool(object) to include your new tool name, class PatronSelect(object).

7. As shown, in the _init_ module, set the label to Bookmobile Patron Selection and the description to "The user selects a possible bookmobile location, and the distance is measured to see how far the first 200 patrons will travel to visit it."

```python
class PatronSelect(object):
    def __init__(self):
        """Define the tool (tool name is the name of the class)."""
        self.label = "Bookmobile Patron Selection"
        self.description = "The user selects a possible bookmobile location, and the distance is measured" + \
        " to see how far the first 200 patrons will travel to visit it."
        self.canRunInBackground = False
```

There are no parameters for this tool because the input is a shapefile that the head librarian already created. Move directly to adding the code to the execute module.

8. Copy the code from the text file starting with the line "# Set up the environment" down to "# Move to the next site and repeat." Paste the code to the execute module right after the comment "The source code of the tool." Be careful to match the indent levels of the existing tool with the new code. Confirm that the indent levels for the for, while, and if statements have been maintained.

```python
def execute(self, parameters, messages):
    """The source code of the tool."""
    # Set up the environment.
    arcpy.env.workspace = r"C:\EsriPress\GISTPython\Data\City of Oleander.gdb\\"
    arcpy.env.overwriteOutput = True

    # Set up cursor for the bookmobile sites.
    arcpy.MakeFeatureLayer_management("Parcels","Parcels_Lyr",'"DU"= 1')
    arcpy.MakeFeatureLayer_management("BookmobileLocations","Locations_lyr")

    siteCursor = arcpy.da.SearchCursor("Locations_lyr","Marker")
    for row in siteCursor:
        siteName = row[0]

        # Select parcels within 150 feet of the new selection.
        arcpy.Select_analysis("Locations_lyr",\
        r"C:\EsriPress\GISTPython\MyExercises\Scratch\Temporary Storage.gdb\SiteTemp",\
        '"Marker" = \'' + siteName + "\'")

        arcpy.SelectLayerByLocation_management("Parcels_lyr","WITHIN_A_DISTANCE",\
        r"C:\EsriPress\GISTPython\MyExercises\Scratch\Temporary Storage.gdb\SiteTemp",\
        "150","NEW_SELECTION")

        # Start a while statement until number of dwelling units exceeds 200.
        parcelCount = int(arcpy.GetCount_management("Parcels_lyr").getOutput(0))
        # print parcelCount - replace in Python toolbox.
        arcpy.AddWarning(parcelCount)

        myCount = 1
        while parcelCount < 200:
            # All statements at this indent level are part of the while loop.

            # Add to the selected set all property within 150 feet, and redo count.
            # SelectLayerByLocation_management (in_layer, overlap_type, select_features,
            # search_distance, selection_type).
            arcpy.SelectLayerByLocation_management("Parcels_lyr","WITHIN_A_DISTANCE",\
            "Parcels_lyr","150","ADD_TO_SELECTION")

            parcelCount = int(arcpy.GetCount_management("Parcels_lyr").getOutput(0))
            # print parcelCount - replace in Python toolbox.
            arcpy.AddWarning(parcelCount)
            if myCount == 8:
                parcelCount = 200
            else:
                myCount = myCount + 1

            # Exit the while statement when count exceeds 200.

        # Export the selected features to a new feature class using the Marker name.
        arcpy.CopyFeatures_management("Parcels_lyr", r"C:\EsriPress\GISTPython\MyExercises\MyAnswers.gdb\\"\
        + siteName.replace(" ","_"))
        # print siteName + " Output OK!" - replace in Python toolbox.
        arcpy.AddWarning(siteName + " Output OK!")
        # Move to the next site, and repeat.
    return
```

The code will work as written, except for the print statements. These are fine for stand-alone Python scripts, but once you move into the ArcGIS environment, you must change these to an ArcPy message command.

9. Scroll through the code, and make these three replacements of the print command, as shown:

```
# print parcelCount
• arcpy.AddWarning(parcelCount)
#print parcelCount
• arcpy.AddWarning(parcelCount)
#print siteName + " Output OK!"
• arcpy.AddWarning(siteName + " Output OK!")
```

10. Save and close the script. Right-click MyExercises and click Refresh to update the code. Your Python toolbox should now look like this:

```
☐ 🗀 C:\EsriPress\GISTPython
    ⊞ 🗀 Data
    ⊞ 🗀 Maps
    ☐ 🗀 MyExercises
        ⊞ 🗀 Scratch
        ☐ 🗊 MyAnswers.gdb
            ☐ 🕮 Proximity
                🗊 Bookmobile Patron Selection
```

11. Double-click the Bookmobile Patron Selection tool to run it. There are no input parameters, so click OK at the prompt.

The tool should run exactly as it did as a stand-alone script and produce the patron counts as expected.

Exercise 4-1

Open the map document Exercise 4-1. Edit the code file for the Proximity Tools toolbox. Turn the stand-alone script you wrote for exercise 2-4 into a Python toolbox tool in the existing toolbox. You will need to add the tool to the Toolbox class, then duplicate the PatronSelect tool class, and modify the tool to create the Exercise 2-4 tool class. Be careful with the indentation and variable types, and be sure to replace any Python-only commands, such as print, with ArcPy commands.

Tutorial 4-1 review

The Python toolbox contains and runs code in exactly the same way as a Python script tool, but the interface is much easier to set up. A single .pyt file will hold all the necessary parameters for all the tools that will appear in the toolbox. These parameters can be controlled in exactly the same way as a script tool, but because everything is in a single file, .pyt files are much easier to share with others. A user can receive a .pyt file and place it in any folder. When the file is accessed in the Catalog window, it will appear exactly like any other toolbox without any further action from the user.

Each tool that you add to the toolbox gets its own class, and each class contains separate functions. The various functions create and manage the self object, which is used to store values associated with the function. Within the class, these values can be passed freely from one function to another, allowing one function to interact with or control other functions.

Study questions

1. What programs are able to edit a .pyt file?
2. Can tools be moved from the Python toolbox to custom toolbars? Try this maneuver, or research its possibility in ArcGIS for Desktop Help.
3. What are some of the drawbacks of Python toolboxes?

Tutorial 4-2 Setting up value validation

One of the best things you can have your script do is to validate user input, which may keep users from typing incorrect values that may cause your scripts to crash.

Learning objectives

- Accepting user input
- Setting up data validation
- Formatting Python toolbox messages

Preparation

Research the following topics in ArcGIS for Desktop Help:

- "The Python toolbox template"
- "Defining a tool in a Python toolbox"
- "Accessing parameters within a Python toolbox"
- "Defining parameters in a Python toolbox"

Introduction

Turning a Python script tool into a Python toolbox can get a little more complicated when the script contains user-defined parameters. To define parameters in the stand-alone script, you first have to write the code to accept values into a variable, which involves the use of the command arcpy.GetParameterAsText(), or for outputs, use arcpy.SetParameterAsText(). Then for each defined parameter, you must define a list of properties when you create the script tool in a toolbox. These properties can be altered but require the juggling of two files and coordinated saves to make them work.

As described in the introduction to this chapter, the Python toolbox tools have all their parameters defined within the Python script. Editing becomes faster and easier, and there are more options than in a script tool. When moving code from a stand-alone script or script tool to the Python toolbox, you should look through the code and find occurrences of the ArcPy parameter commands. These commands will be moved into the parameters module of the Python toolbox code along with any necessary filters or settings.

Scenario

In tutorial 2-5, you wrote an application to use the selected feature in the Fire Department's box zone map to summarize the buildings it contained. In tutorial 2-6, you made the tool interactive by adding a user interface to accept the box number and the building types to summarize. The input for building types had an associated value list to ensure that the user selected an appropriate code, with a chart of the building type descriptions then displayed in the context Help.

For this tutorial, take the code you developed for tutorial 2-6, and turn it into a Python toolbox tool. You may also want to add validation code to ensure that the box number the user enters is valid.

Data

A completed version of the code is in text file Tutorial 4-2 in the Data folder, so your task is to move the code into the Python toolbox structure to make it more portable. The items to be selected are the building footprints for all of Oleander, which are stored in the feature class BldgFootprints in the Planimetrics folder in the City of Oleander geodatabase. The feature class has a field named UseCode, which contains a code for each building type. The following list contains the codes for the building types:

1 = Single Family
2 = Multi-Family
3 = Commercial
4 = Industrial
5 = City Property
6 = Storage Sheds
7 = Schools
8 = Church

The data being used for the selections is a set of polygon feature classes of box zones in the FireBoxMaps feature dataset in the City of Oleander geodatabase. Looking at these feature classes, you will see that there are a large number of these files, one for each box zone. Note also that the numbers are not sequential. Some ranges are skipped, and some box zones appear in two files, which will affect the valid list of choices when you build the data integrity rules.

SCRIPTING TECHNIQUES

You learned in this chapter's introduction how to build a value list for a parameter, and in this tutorial, you will build a value list for the building type codes. Set that parameter's filter type property to Value List, and define a filter list object containing the choices.

You will also create custom Help for this tool. The user will need to know what the building codes mean to be able to make appropriate choices.

Add some code to validate the box number entry. There are too many box numbers to make a usable value list, but it would be a good data integrity practice to make sure the number entered by the user is an actual box number. Verifying the box number is done in a special validation area of the Python toolbox code—the updateParameters module in each tool class. Commands in this area are called anytime an input parameter is changed. The simple validation is to ensure the number entered for the box number is within the allowable range. If the number is not within this range, an error message can be generated, and the script can be held until a proper number is entered.

Set up value validation

1. Open the map document Tutorial 4-2 and text file Tutorial 4-2 in the Data folder. Examine the script from tutorial 2-6, as shown:

```python
# Import the modules
import arcpy

# Set up the environment
arcpy.env.workspace = r"C:\EsriPress\GISTPython\Data\City of Oleander.gdb"

# Get input from the user
#    The first will be the box number to act upon - index 0
#    The second will be the building type to count - index 1
boxNumber = arcpy.GetParameterAsText(0)
buildingType = arcpy.GetParameterAsText(1)


# Make feature layers from the user input
boxLayer = arcpy.MakeFeatureLayer_management(r"C:\EsriPress\GISTPython\Data\City of Oleander.gdb\FireBoxMaps\FireBoxMap_" \
+ str(boxNumber))
buildLayer = arcpy.MakeFeatureLayer_management(r"C:\EsriPress\GISTPython\Data\City of Oleander.gdb\Planimetrics\BldgFootprints", \
"\"UseCode\" = '" + str(buildingType) + "'")
# Use the specified file of box zone to select specified type of building
arcpy.SelectLayerByLocation_management(buildLayer, "HAVE_THEIR_CENTER_IN", boxLayer,"","SUBSET_SELECTION")

# Count the selected features
bldgCount = int(arcpy.GetCount_management(buildLayer).getOutput(0))

# Display the results in the geoprocessing Results window
arcpy.AddMessage("The count of buildings is " + str(bldgCount) + ".")

# Create a field to store the results

# 1 = Single Family (SFCount)
# 2 = Multi-Family (MFCount)
# 3 = Commercial (ComCount)
# 4 = Industrial (IndCount)
# 5 = City Property (CityCount)
# 6 = Storage Sheds (ShedCount)
# 7 = Schools (SchCount)
# 8 = Church (ChurCount)

if buildingType == 1:
    newField = "SFCount"
elif buildingType == 2:
    newField = "MFCount"
elif buildingType == 3:
    newField = "ComCount"
elif buildingType == 4:
    newField = "IndCount"
elif buildingType == 5:
    newField = "CityCount"
elif buildingType == 6:
    newField = "ShedCount"
elif buildingType == 7:
    newField = "SchCount"
else:
    newField = "ChurCount"

arcpy.AddField_management(boxLayer,newField,"LONG")

# Store the results in the field
arcpy.CalculateField_management(boxLayer,newField,bldgCount)
```

2.   **Right-click your MyExercises folder, and create a new Python toolbox named**
Fire Department.pyt. **Open the toolbox for editing.**

3.   **In the Toolbox class, set the label to** Fire Department Tools **and the alias to** fdept. **Then replace
the default value Tool with** BoxBldgCount, **as shown:**

```
class Toolbox(object):
    def __init__(self):
        """Define the toolbox (the name of the toolbox is the name of the .pyt file)."""
        self.label = "Fire Department Tools"
        self.alias = "fdept"

        # List of tool classes associated with this toolbox
        self.tools = [BoxBldgCount]
```

4.   **Change the name of the Tool class to** class BoxBldgCount(object).

5.   **In the _init_ module of the BoxBldgCount class, change the label to** Box Zone Building Count **and
the description as shown:**

```
class BoxBldgCount(object):
    def __init__(self):
        """Define the tool (tool name is the name of the class)."""
        self.label = "Box Zone Building Count"
        self.description = "The user will enter a box number and a building type." + \
        " The tool will add a field and store the building count for the selected type."
        self.canRunInBackground = False
```

Next set up the two input parameters. The first is simply a request for the user to enter a number
(integer). Look up the keyword for this data type in ArcGIS for Desktop Help.

6.   **Add the code to accept the box number from the user, and set up the other properties of the
input parameter to match the original application, as shown:**

```
    def getParameterInfo(self):
        """Define parameter definitions"""
        param0 = arcpy.Parameter(
        displayName="Enter the Box Number",
        name="inBox",
        datatype="GPLong",
        parameterType="Required",
        direction="Input")
```

The numbers entered into this parameter must correspond to the box zone numbers. Without
this number validation, the user could enter a number that would cause an error in the script. The
validation code is written in the updateMessages module. The updateMessages module is used
to provide feedback to the user based on the values they enter as parameters and is displayed
to the user before they click OK to run the tool. There is an important distinction here. The
updateParameters module validates user input as it is typed, and the updateMessages module can
return feedback to the user based on the parameters entered.

The valid box zone numbers are from 100 to 117, 200 to 210, 300 to 309, and 318 to 321. An if statement can check the validity of the box numbers, and if the number is invalid, a script message can be sent to the input dialog box.

The two most common script messages are setWarningMessage(), which displays a yellow triangle icon on the input dialog box, and setErrorMessage(), which displays a red X on the input dialog box. The error message will also stop the script from running until the error is resolved.

7. **Scroll to the updateMessages module in the tool's class. Add an if statement to validate that the box number entered is within the acceptable range. Note that the range tool is not inclusive of the second value. Also add a setErrorMessage() command to stop the script if the number fails the validation, as shown:**

```
def updateMessages(self, parameters):
    """Modify the messages created by internal validation for each tool
    parameter.  This method is called after internal validation."""
    if parameters[0].value in range(100,118):
        arcpy.AddMessage("Value is OK.")
    else:
        parameters[0].setErrorMessage(str("This number is out of range"))
    return
```

Your turn

The last set of code validates the box maps in District 1. Using the ranges stated previously, add the code to validate the other acceptable ranges and to provide an applicable error message, as shown:

```
def updateMessages(self, parameters):
    """Modify the messages created by internal validation for each tool
    parameter.  This method is called after internal validation."""
    if parameters[0].value in range(100,118):
        arcpy.AddMessage("Value is OK.")
    elif parameters[0].value in range(200,211):
        arcpy.AddMessage("Value is OK.")
    elif parameters[0].value in range(300,310):
        arcpy.AddMessage("Value is OK.")
    elif parameters[0].value in range(318,322):
        arcpy.AddMessage("Value is OK.")
    else:
        parameters[0].setErrorMessage(str("This number is out of range"))
    return
```

The entry and validation of the first parameter is complete. Next, add and configure a value list parameter for the building type. The entry portion of the parameter is basically the same as the first parameter. Give the user a list of choices using the descriptions by setting the filter type and the filter list. Then include code to set the return value based on the user's choice.

8. Set up a second parameter (param1) with the display name Select a Building Type and the name bldgType. The data type is a string, and the keyword for that data type is found in ArcGIS for Desktop Help.

```
param1 = arcpy.Parameter(
displayName="Select a Building Type",
name="bldgType",
datatype="GPString",
parameterType="Required",
direction="Input")
```

9. Now add a filter to the parameter to make it a value list with the text descriptions of the building types, as shown:

```
param1.filter.type = "ValueList"
param1.filter.list = ["Single Family","Multi-Family","Commercial","Industrial","City Property", \
                      "Storage Sheds","Schools","Church"]
```

10. After defining the two parameters, add them to the parameters list returned by the getParameterInfo method, as shown:

```
params = [param0,param1]
return params
```

11. Save and close the Python script. Right-click your MyExercises folder, click Refresh to ensure that the Python toolbox loads the most recent code, and then double-click the Box Zone Building Count tool to run it. Note the inputs and validations. Try several box numbers (press Tab after typing each one), and examine the drop-down lists, as shown. When you are finished, click OK or Cancel. The script has no code in the execute module, so nothing will happen, but you can see how the interface operates.

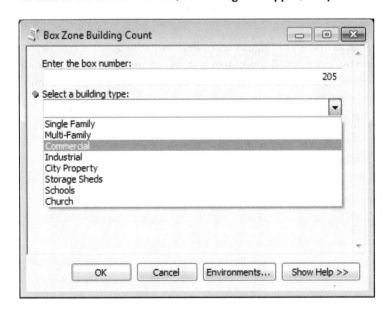

Now add all the executable code to run the application.

12. Start editing the Python toolbox. Copy the code from the text file Tutorial 4-2 in the Data folder, starting with the line "# Get input from the user" through to the end. Paste the code to the execute module above the return statement, and correct the indent levels as necessary. Note the code to accept user input, as shown:

```
def execute(self, parameters, messages):
    """The source code of the tool."""
    # Import the modules
    import arcpy


    # Set up the environment
    arcpy.env.workspace = r"C:\EsriPress\GISTPython\Data\City of Oleander.gdb"

    # Get input from the user
    #     The first will be the box number to act upon - index 0
    #     The second will be the building type to count - index 1
    boxNumber = arcpy.GetParameterAsText(0)
    buildingType = arcpy.GetParameterAsText(1)
```

The original script used arcpy.GetParameterAsText() to gain input from the user, but the Python toolbox uses parameters from the getParameterInfo method. To add the correct input for the executable code, change the arcpy.GetParameterAsText() code to parameters[] code, making sure to keep the same index numbers.

13. Change the data acceptance code to use the parameters from the toolbox script, as shown:

```
    # Get input from the user
    #     The first will be the box number to act upon - index 0
    #     The second will be the building type to count - index 1
    boxNumber = parameters[0].value
    buildingType = parameters[1].value
```

There is one other change to make. The original script dealt with the building types as integers, and now they are being selected as strings. This problem is easily repaired by changing the values to integers based on their text description. However, when you make this change, you must cast this variable to a string before you can use it in any message.

14. In the executable code just above the line "# Make feature layers . . . ," add a set of if-elif-else statements to convert the building type description into an integer using the values from the list provided in the data description at the start of this tutorial, as shown:

```
# Convert the string descriptions into integers
# "Single Family","Multi-Family","Commercial","Industrial","City Property",
# "Storage Sheds","Schools","Church"
if buildingType == "Single Family":
    buildingType = 1
    arcpy.AddMessage("You selected Single Family.")
elif buildingType == "Multi-Family":
    buildingType = 2
    arcpy.AddMessage("You selected Multi-Family.")
elif buildingType == "Commercial":
    buildingType = 3
    arcpy.AddMessage("You selected Commercial.")
elif buildingType == "Industrial":
    buildingType = 4
    arcpy.AddMessage("You selected Industrial.")
elif buildingType == "City Property":
    buildingType = 5
    arcpy.AddMessage("You selected City Property.")
elif buildingType == "Storage Sheds":
    buildingType = 6
    arcpy.AddMessage("You selected Storage Sheds.")
elif buildingType == "Schools":
    buildingType = 7
    arcpy.AddMessage("You selected Schools.")
else:
    buildingType = 8
    arcpy.AddMessage("You selected Church.")
# Make feature layers from the user input
```

15. Save and close the script. Right-click your MyExercises folder, and click Refresh to update the code. Double-click the Box Zone Building Count tool to run it. Get a count of storage sheds in box 202, as shown:

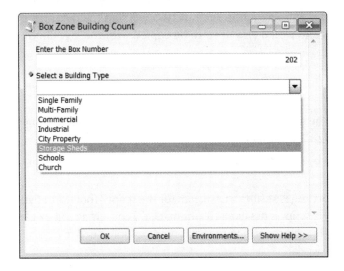

Try this with other combinations of box numbers and building types. Afterward, inspect the attribute tables for the input files, and note the addition of the fields and counts to the tables.

Exercise 4-2

The Fire Department has been using the lane-miles tool that you wrote in exercise 2-5 and wants to share it with other departments in the region. It would be easier to send around a .pyt file than having to send a .tbx file, a .py script file, and instructions on setting up the files. The fire chief has asked that you make a Python toolbox tool for the lane-miles calculations.

Review the code you wrote for exercise 2-5, and use that code as the basis for the new toolbox. Add any input validation that you feel might be warranted. As a bonus, add the tool to the existing toolbox, and put the toolbox in a toolset named **Street Tools**.

Tutorial 4-2 review

This tutorial includes the addition of input parameters for the tools. In the script tools, the user input has to be configured separately when the script is added to the custom toolbox. The Python toolbox contains a special method for getting user input. All the parameters that are set in the script tool parameters are stored in the same file as the executable code.

Many of the same validation controls used in script tools are also available in the Python toolbox. This tutorial includes a value validation for the box number and a value list filter to create a list of choices for the building type. The validation code is added to a function in the tool class, and then the behavior is added to the tool and activated when it runs.

Documentation can be added to the tools in a Python toolbox. Simply right-click the tool and click Item Description > Edit. This opens a dialog box to build the same type of context-sensitive Help that you can build with a script tool.

The convenience of the Python toolbox is that all the code is easily contained in a single file. As a result, the toolbox and its custom tools are easier to share with others. However, if you choose to make custom Help files for a Python toolbox, you will find that all the help is written to separate XML files and would also need to be copied with the .pyt file when being shared.

Study questions

1. What did you find to be the easiest and the most difficult things about building a Python toolbox as compared to a script tool?
2. What changes had to be made to the code to move a tool from a custom toolbox to a Python toolbox?
3. Search the Esri Support pages at support.esri.com for the script "Toolbox to Python Toolbox Wrapper." This script is designed to automatically convert a toolbox file to a Python toolbox file. Test this script, and see what modifications need to be made to the code for this script to work.

Tutorial 4-3 Setting up dependencies

Python toolboxes allow the programmer to make decisions on input values based on other choices the user may make. These types of dependencies make the tool user-friendly and can prevent errors in data entry.

Learning objectives

- Setting Python toolbox parameters
- Using if-elif-else logic for input validation
- Designing a user input box

Preparation

Research the following topics in ArcGIS for Desktop Help:

- "Customizing tool behavior in a Python toolbox"
- "Writing messages in a Python toolbox"

Introduction

Another useful feature of the Python toolbox is the validation module updateParameters. This module allows you to control the value of a parameter based on the selected input of another parameter. The user can select a value, and the script will detect that the value has been changed. The script will then run a set of validation code that can control the value of other parameters based on the first parameter. For instance, you might have a user select a workspace as the first parameter, and restrict the second parameter to showing only the files in that workspace.

This type of validation code can also be used to restrict selections based on another input. Another example would be to have the user select a feature class, and then restrict a second parameter to present only the fields from within that feature class. Using any of these types of value validation will add data integrity rules to your application, which will make the application easier to use and help prevent data input errors.

Scenario

The map generator you made for the city planner in tutorial 3-1 is a perfect candidate for a Python toolbox. The script tool prompted the planner to select a map title from a value list and enter a date. Then the subtitle and map description were populated with prewritten text based on his choice of map title using value validation code. That code was held in a separate area of the script tool parameters, but in the Python toolbox it will be incorporated into the single .pyt file using the updateParameters module. As before, examine the application, and move it to a new Python toolbox.

Data

A completed copy of the code is in text file Tutorial 4-3 in the Data folder, so your task is to move it into the Python toolbox structure to make it more portable.

SCRIPTING TECHNIQUES

The updateParameters method runs each time the script detects that one of the input values has been changed. One example may be to have a description value populated automatically based on the selection of a building type. The first parameter would define a value list of building types for the user to choose from. The value list would help prevent typing mistakes or the entry of a building type that does not exist. The second parameter is simply a text field to hold the description of that building type, as shown:

```
def getParameterInfo(self):
    """Define parameter definitions"""
    param0 = arcpy.Parameter(
    displayName="Select a Building Type",
    name="bldgType",
    datatype="GPString",
    parameterType="Required",
    direction="Input")

    param0.filter.type = "ValueList"
    param0.filter.list = ["Single Family","Multi-Family","Commercial","Industrial","City Property", \
                          "Storage Sheds","Schools","Church"]

    param1 = arcpy.Parameter(
    displayName="Description",
    name="bldgDesc",
    datatype="GPString",
    parameterType="Required",
    direction="Input")

    params = [param0,param1]
    return params
```

In the updateParameters module, add code to determine what building type was chosen, and set the description to a preset value. In this example, an if statement is used to check all the possible values. Notice that the parameter reference is now parameters[0] instead of param0 because the values have been passed from one module to another. The square brackets are used to contain the index number of a value in the parameters object. Note also that there is an inclusion of a null or empty ("") value in the else statement, as shown in the graphic. This empty value will be used to set the value of the parameter to null when the script is initialized.

```
def updateParameters(self, parameters):
    """Modify the values and properties of parameters before internal
    validation is performed.  This method is called whenever a parameter
    has been changed."""
    # Set a default value for the description based on the building type selected
    if parameters[0].value == "Single Family":
        parameters[1].value = "Single family detached dwellings at 3 units per acre."
    elif parameters[0].value == "Multi-Family":
        parameters[1].value = "Apartments and Townhouse dwellings at up to 25 units per acre."
    elif parameters[0].value == "Commercial":
        parameters[1].value = "Neighborhood and community business districts"
    elif parameters[0].value == "Industrial":
        parameters[1].value = "Industrial and manufacturing located in appropriate areas."
    elif parameters[0].value == "City Property":
        parameters[1].value = "Land owned and maintained by the City of Oleander."
    elif parameters[0].value == "Storage Sheds":
        parameters[1].value = "Storage sheds and detached garages."
    elif parameters[0].value == "Schools":
        parameters[1].value = "Primary and secondary school locations."
    elif parameters[0].value == "Church":
        parameters[1].value = "Churches and places of worship with appropriate permits."
    else:
        parameters[1].value = ""
    return
```

When the tool runs, the user sees the dialog box on the left. When the selection of a
building type is made, the validation code automatically runs, detects the choice,
and fills in the description as shown on the right:

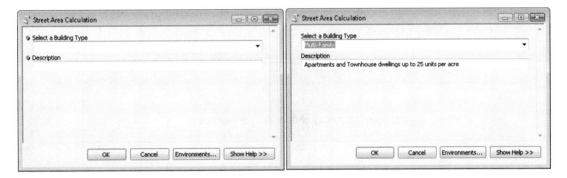

You may also want to control the choices of a value based on the first value. In this
second example, the user is asked in the first parameter to select a feature class.
Then a list of fields is presented in the second parameter based on the chosen
feature class. The first parameter is set up like normal, as shown:

```
# First parameter
param0 = arcpy.Parameter(
    displayName="Input Features",
    name="in_features",
    datatype="DEFeatureClass",
    parameterType="Required",
    direction="Input")
```

The second parameter includes the parameterDependencies property, which sets the choices based on the first parameter. Note that this is a property of the second parameter, but it references the first parameter as the dependency, as shown:

```
# Second parameter
param1 = arcpy.Parameter(
    displayName="Storage Field",
    name="storField",
    datatype="Field",
    parameterType="Required",
    direction="Input")
# Only the fields from the param0 feature class will be shown
param1.parameterDependencies = [param0.name]
```

This property can be set either in the getParameterInfo method or the updateParameters method, but remember that if you use the updateParameters method, you must use the global parameters[] list object.

The input dialog box lets the user navigate to a feature class, and then the second parameter is populated with a list of field names dependent on the selected feature class. This example shows the user selecting a feature class as the first input parameter. A list of field names is then retrieved from the feature class and displayed as a value list for the second input parameter. The list of fields is dependent on the choice of feature class, as shown:

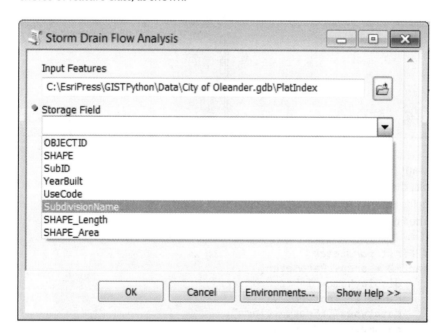

Notice that the parameterDependencies property references only the name property of the first parameter, so how does it know what values to use in the dependency? The parameterDependencies property checks the data type of the parameter. The first parameter has a data type of feature class, and the second parameter has a data type of field. By having a dependency set, the second parameter gets a list of fields from the feature class selected in the first parameter. For example, if the first parameter asked for a file name, the dependency would be a list of files.

Controlling user input through lists of valid choices builds strong data integrity rules that will make your tools more user-friendly.

Set up dependencies

1. **Open the map document Tutorial 4-3.**

A map of Oleander is displayed with three layers symbolized to show property values, land use, and the year of construction. As in tutorial 3-1, you will write a Python toolbox tool that will open a dialog box and allow the user to select a map type and enter a title and date. Then the other text items will be populated and the correct layer turned on to create the requested map.

2. **Open the text file Tutorial 4-3 from the Data folder.**

This file contains the code from tutorial 3-1, and in a special section at the bottom, you will find the validation code. This code was in a separate file in the script tool, and it controls some of the text entry.

3. **In your MyExercises folder, create a new Python toolbox named** Planning Maps.pyt, **and open the toolbox for editing.**

4. **In the Toolbox class, change the label to** Planning Tools **and the alias to** pltools. **This toolbox will have only one tool. Name it** makeMaps, **as shown:**

```
class Toolbox(object):
    def __init__(self):
        """Define the toolbox (the name of the toolbox is the name of the .pyt file)."""
        self.label = "Planning Tools"
        self.alias = "pltools"

        # List of tool classes associated with this toolbox
        self.tools = [makeMaps]
```

Now set up the tool's properties.

5. **Change the Tool class to** makeMaps, **set the label to** Make Planning Maps, **and type an appropriate description, as shown:**

```
class makeMaps(object):
    def __init__(self):
        """Define the tool (tool name is the name of the class)."""
        self.label = "Make Planning Maps"
        self.description = "The user can select between three maps that will then be automatically " + \
                    "exported to a PDF file."
        self.canRunInBackground = False
```

Next define the input parameters. Review the code from the script tool, and note that there are four inputs: map title, subtitle, date, and description, as shown:

```
# Create variables to accept user input for the map title,
# map subtitle, current date, and description.
newTitle = arcpy.GetParameterAsText(0)
newSubTitle = arcpy.GetParameterAsText(1)
newDate = arcpy.GetParameterAsText(2)
newDesc = arcpy.GetParameterAsText(3)
```

6. Add the code to create four input parameters with the same name and index numbers as the original script. Set the map title as a value list with the choices of Property Value, Land Use, or Year of Construction. **Work on writing this code yourself first, and then check it against the code block, as shown:**

```
def getParameterInfo(self):
    """Define parameter definitions"""
    param0 = arcpy.Parameter(
    displayName="Enter a new Map Title",
    name="newTitle",
    datatype="GPString",
    parameterType="Required",
    direction="Input")

    param0.filter.type = "Value List"
    param0.filter.list = ["Property Value","Land Use","Year of Construction"]

    param1 = arcpy.Parameter(
    displayName="Enter a new Map Subtitle",
    name="newSubTitle",
    datatype="GPString",
    parameterType="Required",
    direction="Input")

    param2 = arcpy.Parameter(
    displayName="Enter a Date",
    name="newDate",
    datatype="GPString",
    parameterType="Required",
    direction="Input")

    param3 = arcpy.Parameter(
    displayName="Enter a Map Description",
    name="newDesc",
    datatype="GPString",
    parameterType="Required",
    direction="Input")

    params = [param0,param1,param2,param3]
    return params
```

7. Save and close the script. Right-click your MyExercises folder, and click Refresh to update the code. Double-click the script tool Make Planning Maps to run it. If constructed correctly, your input dialog should look like the graphic, with a list of choices for the map type.

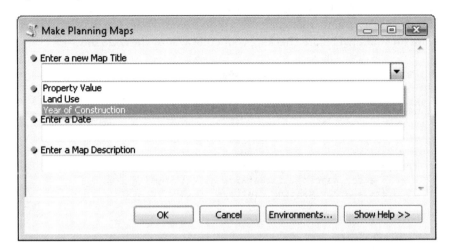

The inputs for subtitle and description come from prewritten text in the original script tool's parameters. The code to perform this task is in the updateParameters method of the tool's class. Remember that the updateMessages method validates an entry and stops the script if the entry is not valid, and the updateParameters method controls the value of parameters based on the values of other parameters.

8. Edit the Planning Maps toolbox. Copy the validation code at the bottom of the text file Tutorial 4-3, and paste the code to the updateParameters code block. Be careful to set the indent levels correctly, as shown:

```
def updateParameters(self, parameters):
    """Modify the values and properties of parameters before internal
    validation is performed.  This method is called whenever a parameter
    has been changed."""
    if parameters[0].altered:
        if parameters[0].value == "Property Value":
            parameters[1].value = "Derived from Appraisal District files"
            parameters[3].value = "The property values for Oleander are appraised and determined by the " + \
            "Tarrant County Appraisal District. Monthly reports are used to update the parcel database to " + \
            "keep the data as current as possible. Property shown with a value of $0.00 may be under protest " + \
            "and the new value will be available after the owner has had a public hearing to settle the matter."
        elif parameters[0].value == "Land Use":
            parameters[1].value = "Current land use for all property"
            parameters[3].value = "Land use is determined by city staff to fit into one of many categories. " + \
            "The initial land use is determined by the occupant's Certificate of Occupancy application, and " + \
            "later field verified. Any changes in land use are noted on an annual basis, which may trigger " + \
            "the occupant to obtain a new Certificate of Occupancy or if necessary a Specific User Permit."
        else:
            parameters[1].value = "Historical construction data shown by decade"
            parameters[3].value = "The records of construction for Oleander date back to the 1940s when " + \
            "Oleander was first incorporated. Building constructed prior to that are grouped into the pre-1940s " + \
            "category since the city does not have verification of dates earlier than that. Vacant property and property " + \
            "where no construction date is known are shown with no color shading. The purpose of this map is to " + \
            "give an overview of the growth of Oleander over the years and not to determine a specific building's " + \
            "date of construction."
    return
```

Note that this code begins with an if statement and uses the .altered property to determine whether the user made a selection from the value list. Normally, this code would be run on initialization, but the .altered property means that the code will run only when the map title parameter has been altered (and not on initialization).

The final step is to paste the executable code. Make sure to read through the code and make any necessary changes as discussed before, such as altering the commands to accept user input.

9. Copy and paste the executable code from the text file Tutorial 4-3 to the appropriate section of the .pyt file. Try to determine what code to copy and what changes to make to object names and indentations on your own before checking your code against the code shown:

```python
def execute(self, parameters, messages):
    """The source code of the tool."""

    # Create a map document object for the currently open map document;
    # a data frame list of the only data frame in this map document,
    # Note: leave off the index [0] to get all data frames;
    # a list of layers in this map document
    thisMap = arcpy.mapping.MapDocument("CURRENT")
    myDF = arcpy.mapping.ListDataFrames(thisMap)[0]
    myLayers = arcpy.mapping.ListLayers(myDF)

    # Create variables to accept user input for the map title,
    # map subtitle, current date, and description.
    newTitle = parameters[0].value
    newSubTitle = parameters[1].value
    newDate = parameters[2].value
    newDesc = parameters[3].value

    # Create a list of the map elements
    # These will be accessed by their names
    myElements = arcpy.mapping.ListLayoutElements(thisMap, "TEXT_ELEMENT", "map*")

    # Scroll through the list of elements with a for statement
    # Use an if statement to test for each of the element names you
    # are changing and replace the text with the user input
    for element in myElements:
        if element.name == "Map Title":
            element.text = newTitle
        elif element.name == "Map Subtitle":
            element.text = newSubTitle
        elif element.name == "Map Date":
            element.text = newDate
        elif element.name == "Map Description":
            element.text = newDesc

    # Refresh the map so that the changes can be seen
    arcpy.RefreshActiveView()

    # Determine which title has been chosen
    # Use the title name to determine which layers to make visible
    if newTitle == "Property Value":
        for layer in myLayers:
            if layer.name == "Property Value 2012":
                layer.visible = "True"
            elif layer.name == "Land Use":
                layer.visible = "False"
            elif layer.name == "Year of Construction":
                layer.visible = "False"
            else:
```

```
            layer.visible = "False"
        else:
            layer.visible = "True"
elif newTitle == "Land Use":
    for layer in myLayers:
        if layer.name == "Property Value 2012":
            layer.visible = "False"
        elif layer.name == "Land Use":
            layer.visible = "True"
        elif layer.name == "Year of Construction":
            layer.visible = "False"
        else:
            layer.visible = "True"
else:
    for layer in myLayers:
        if layer.name == "Property Value 2012":
            layer.visible = "False"
        elif layer.name == "Land Use":
            layer.visible = "False"
        elif layer.name == "Year of Construction":
            layer.visible = "True"
        else:
            layer.visible = "True"

# Refresh the map so that the changes can be seen
arcpy.RefreshActiveView()

# Output completed map to a PDF file
arcpy.mapping.ExportToPDF(thisMap,r"C:\EsriPress\GISTPython\MyExercises\\" + \
newTitle.replace(" ","_") + "_" + newDate.replace(" ","_"))

# Clear this map document from memory so that it won't be locked
del thisMap
return
```

10. Save and close the .pyt file. Right-click MyExercises, and click Refresh to update the code. Run the Make Planning Maps tool, and verify that the tool is making the maps requested.

Exercise 4-3

For this exercise, create a new Python toolbox without the benefit of existing code. The tool will be used to select a workspace and then get the spatial reference of a feature class in that workspace. You should make two input parameters: the first to find a workspace, and the second to make a value list based on the feature classes in that workspace.

Write the pseudo code necessary for the application, and sketch out how the input interface will work. Then build and debug the application. As a bonus, allow multiple feature classes to be selected once the workspace is identified.

Tutorial 4-3 review

Much like the value validation code from a script tool, the Python toolbox allows the user's input to control other input values, which is done with a dependency property. Note that the parameter using a dependency will match the data type of the parameter it references.

The other important technique in this tutorial is the sharing of parameter values across methods in the Python code. As long as the methods are in the same class, values will be passed from one method to the other. In this tutorial, each method in a class brings in the self object, and several of the methods also bring in parameter values. By bringing in the self object and parameter values, you can accept or manipulate values in one function, and have those values available for use in other functions as well.

Study questions

1. Show an example in the code where a value is passed from one function to another.
2. Explain the difference between the updateParameters method and the updateMessages method.
3. Research the isLicensed method, and describe its purpose.

Chapter 5

Python add-ins

Introduction

You can modify your user experience with ArcGIS in many ways, including with the use of ModelBuilder, script tools, and Python toolboxes. You can use any of these applications to create custom tools that can reside on custom menus to make the software easier to run and your data easier to manage. One of the newest ways to create custom applications is with a Python add-in, which is a program that you design, write, and compile yourself. Add-ins allow fully customized menus to be developed, which can contain a variety of tools including ArcGIS system tools. The compiled add-in file can then be easily distributed to other users.

Python add-ins are a bit of work to create, so it is important to know when they are necessary. These add-ins have two main functions that distinguish them from other methods of customization. First, Python add-ins allow you to build tools to interact with the map display directly. Examples of this include having the user select features, draw boxes for zooming, or click a location on the map. In addition to the tools interacting with the map, the add-ins can also interact with each other. One tool can pass values to another tool or even keep another tool disabled until a valid input is made.

The other important function of Python add-ins is the capacity for having the application react to other events. Examples of this function include having the map document automatically save changes when you stop editing, zooming the map to the full extent of a new layer that is added, or automatically closing the Catalog window when a map is printed.

Once your add-ins are built, they are easy to share with others. A single file with an .esriaddin extension contains everything necessary for your application. When you combine it with a map package, it is easy to share the application and data using only two files.

Add-ins for ArcGIS are readily available in other programming languages, such as .NET, XML, and Java. These add-ins were made available in ArcGIS 10.1 for use in ArcMap, ArcScene, ArcGlobe, and ArcCatalog using Python as the development tool, which is the focus of the tutorials in this chapter.

Special introduction: Python add-ins

Many components are available for use in the Python add-ins arena, and this special introduction presents these components and describes how they are implemented. With this knowledge, you will be able to better design your applications.

The first step in creating an add-in is to create the framework using the addin_assistant wizard. This wizard can be downloaded from the Esri Support pages at support.esri.com, and a copy is provided in this book's data in the GISTPython\Python Add-Ins folder. Unzipping the file creates a bin folder that contains the executable file for the wizard, addin_assistant.exe. It is advisable to make a desktop shortcut for this file because it will be used repeatedly during your application development.

Running the addin_assistant.exe file will produce the management screen, or wizard, for your project. Pointing this wizard to an empty folder will create a new add-in, and pointing it to an existing folder that contains an add-in will let you modify that add-in. As an example, a new folder is created to contain an add-in named Water Analysis, as shown:

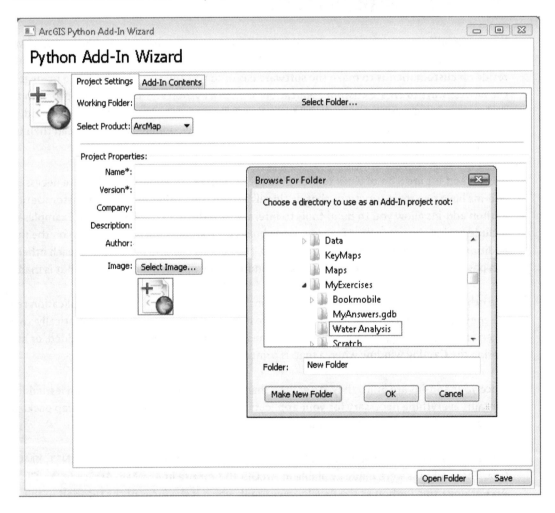

The name of this folder will be displayed as the working folder. Select the ArcGIS module this project will work with. The Select Product drop-down box provides the available choices. This example will be used in ArcMap, as shown:

A name and version number are required. The name is what will appear as the toolbar name in your application. The version is used to track the development of the application. For this example, the name is Water Analysis Tools, and the version is 1.0. Be careful not to provide too long a name because it may not fit the width of your toolbar. The values for company, description, and author are optional and do not affect the running of the application. You can also select a custom image for the project if you like, which is shown in the add-in manager, as shown:

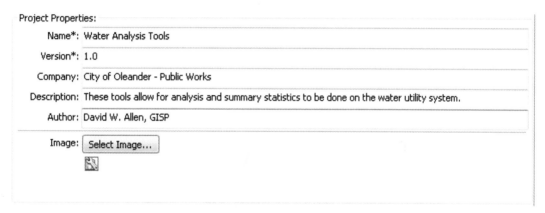

To create the framework in the new folder, click Save. This folder contains several files and some new folders. The new Images folder contains all the icons you assign for use in your application. These are then transmitted with the final .esriaddin file so that other users get the specified look and feel of the add-in. The Install folder holds the Python code for your application, and is where you will go to write your code. The files in the Install folder are created by the add-in manager and should only be edited with extreme caution. As shown in the graphic, the config.xml file is for internal use by the add-in application and controls such things as the order of tools in the toolbox. By default, tools are shown in the order in which you create them, but with cautious editing of the config.xml file, you could rearrange them. The README text file is a generic description of the files associated with the add-in. The makeaddin.py file is the script that actually creates the .esriaddin file for you.

Folders		Name	Date modified	Type	Size
▷ 📁 Scratch	▲	📁 Images	6/21/2013 11:59 AM	File Folder	
▲ 📁 Water Analysis		📁 Install	6/21/2013 11:59 AM	File Folder	
📁 Images		📄 config.xml	6/21/2013 11:59 AM	XML Document	1 KB
📁 Install		📄 makeaddin.py	6/21/2013 11:59 AM	PY File	2 KB
▷ 📁 Python Add-Ins		📄 README.txt	6/21/2013 11:59 AM	Text Document	1 KB

The second tab on the add-in wizard is for the Add-In Contents dialog box. This is where you define the toolbars, menus, and extensions for your project, which in turn create the empty template for these items in a Python script file. For this example, one of each type of content (a toolbar, a menu, and an extension) is added to the Python add-in toolbar. It is important to design the toolbar first because each type of content is added to the add-in script file in the order it was created and will be difficult to move later. As an example, the sketch shown in the graphic was drawn to help design an add-in toolbar containing each of the elements available to you. This design will also be used for the other examples in this introduction.

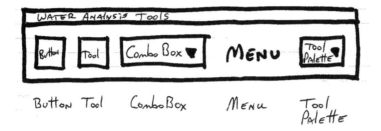

The add-in project begins with creating a toolbar because the toolbar is where all the other components will reside. To create a toolbar, right-click Toolbars in the Python Add-In Wizard dialog box and click New Toolbar. There are just a couple of items to define. The first item is the toolbar's caption, which is what will be displayed along the top of the toolbar. The second item identifies this toolbar uniquely within your Python code. The part of the ID to the right of the period is user definable. If you wanted to programmatically call this toolbar, this ID is what you would use. Select the Show Initially check box to make this toolbox visible automatically when the add-in is loaded, as shown:

Toolbar

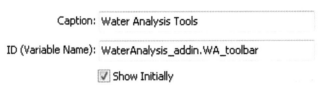

With the toolbar created, the other features can be added to it. For this example, the next thing to add is a button. A button is one of the most straightforward items to add because it basically runs a set of code when the button is pressed. To create a button, right-click the toolbar and click New Button. The selections for other toolbar components are also shown:

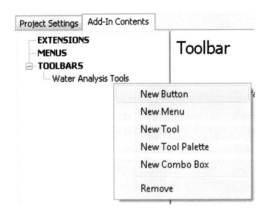

A button, like the other items that go on a toolbar, has many parameters to set and many properties associated with these parameters. The first three parameters are required: the caption, the class name, and the button ID. The caption is the text used for the button's title in the toolbar (if the text is displayed). The caption can contain spaces but should be kept short so that it does not take up too much room on the toolbar. The class name is used to create the section of code, or class, that is accessed when you click the button. This class holds the executable code for whatever task the button performs. The names typically use the up-style naming convention, in which the first letter of each word is capitalized. The class name should describe the tool's processes to help distinguish its use for the programmer. The button ID, like the toolbar ID, is used to identify this tool within the code of your script.

After these required parameters are the Help items. The first pair are the tooltip and message. The tooltip is a set of words or a phrase that will appear when the mouse is paused over the button, and the message is a longer description of the tool's action that appears beneath the tooltip in a smaller font. The next pair are the Help heading and Help content, which appear when the user opens the context-sensitive Help. The Help heading and Help content typically contain the same wording as the tooltip and message unless more information is required beyond these brief descriptions.

You have the option to provide an icon for the button if desired. If an icon is selected, it will be shown in the regular icon size of 16 x 16 pixels to fit the toolbar. The icon will be copied to the images folder within the add-in's folder and included in the compiled add-in file. If an icon is not specified, the button will display the caption text. For this example, the icon is not specified, as shown:

Button

Caption:	Save To Table
Class Name:	Save_Table
ID (Variable Name):	WaterAnalysis_addin.savetable
Tooltip:	Save To Table
Message:	Saves the compiled results to an external tab
Help Heading:	Save To Table
Help Content:	Saves the compiled results to an external tab
Image for control:	

To add the class for the new button to the add-in Python script in the Install folder, click Save. This code contains controls for the button and the executable code for the button's process, as shown:

```python
class Save_Table(object):
    """Implementation for WaterAnalysis_addin.savetable (Button)"""
    def __init__(self):
        self.enabled = True
        self.checked = False
    def onClick(self):
        # Custom executable code for button's task
        pass
```

Note that the class name defined at the button's creation is used for the class name within the code. Although the button ID is not used explicitly in the code, the ID is shown for reference in the purple text in the code in the previous graphic and noted as a button. Remember that only the part of the button ID to the right of the period is necessary to identify the button.

Each class created for a toolbar item returns the object "self." You can use the self object to access the properties that can be set and controlled within this class and the functions that the add-in toolbar will run. Some of the properties and functions of the self object are listed here:

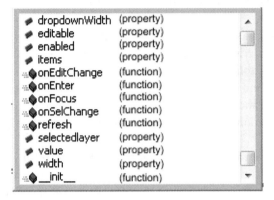

The _init_ function runs when the toolbar is first initialized and can set the properties of the self object. Setting the enabled property to False causes the button to be disabled. The property checked controls the visual appearance of the button either as raised (False) or depressed (True). To control these properties outside this class, reference them as savetable.enabled or savetable.checked, using the button ID.

As an example, if you wanted a button to remain disabled until a second button is pressed, set the enable property of the first button to False and add the enable property to the second button, making sure to reference the first button by its button ID. In this example, the second to last line of code for the second button sets the enable property for the first button to True using the code "<button ID>.enabled = True," as shown:

```
class Save_Table(object):
    """Implementation for WaterAnalysis_addin.savetable (Button)"""
    def __init__(self):
        self.enabled = False
        self.checked = False
    def onClick(self):
        # Custom executable code for button's task
        pass

class Create_Table(object):
    """Implementation for WaterAnalysis_addin.createtable (Button)"""
    def __init__(self):
        self.enabled = True
        self.checked = False
    def onClick(self):
        # Code to create a table
        if exists(table.name):
            savetable.enabled = True
        pass
```

The second function in the button is the onClick action of clicking the button. The onClick function contains code to be executed when the button is pressed.

The project's design diagram shows that the next item to add is a tool. Tools differ from buttons in that they require interaction with the map area when activated. Adding a tool is accomplished by returning to the Python Add-In Wizard, right-clicking the toolbar, and clicking New Tool. The properties for a tool are the same as those for a button, as shown:

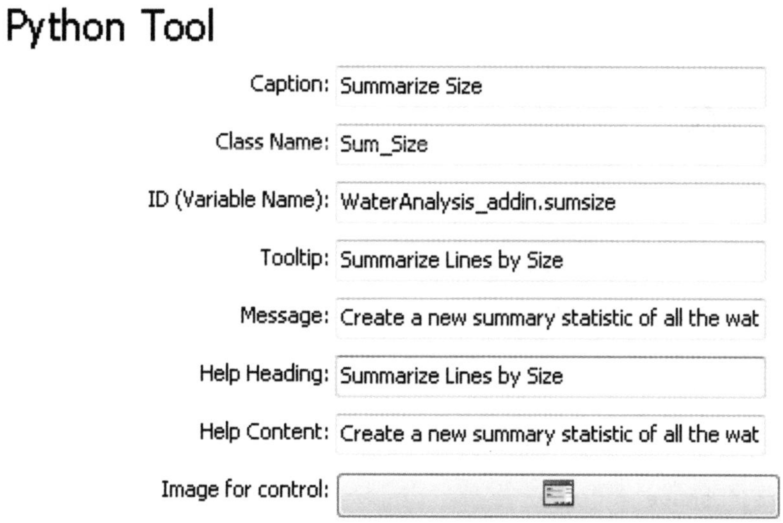

Python Tool

Caption:	Summarize Size
Class Name:	Sum_Size
ID (Variable Name):	WaterAnalysis_addin.sumsize
Tooltip:	Summarize Lines by Size
Message:	Create a new summary statistic of all the wat
Help Heading:	Summarize Lines by Size
Help Content:	Create a new summary statistic of all the wat
Image for control:	

A new class is added to the add-in's script file. The new class includes several functions and allows the control of many properties. The _init_ function and the new class that was created are shown here:

```
class Sum_Size(object):
    """Implementation for WaterAnalysis_addin.sumsize (Tool)"""
    def __init__(self):
        self.cursor = 0
        self.enabled = True
        self.shape = "NONE" # Can set to "Line", "Circle", or "Rectangle" for interactive
                            # shape drawing in the map and to activate the
                            # onLine, onCircle, or onRectangle functions
```

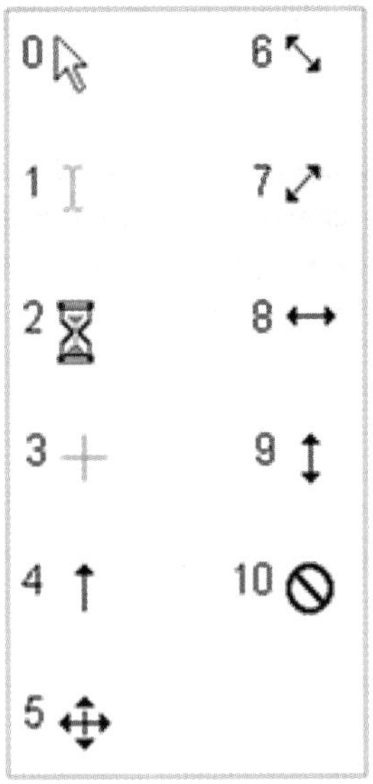

0	6
1	7
2	8
3	9
4	10
5	

The self object has an enable property, which you are already familiar with, and shape and cursor properties. The cursor property sets the icon used for the cursor, examples of which are shown in the graphic, and the shape property determines the geometric shape that will be used to interact with the map (line, circle, or rectangle). When set to None, the interaction is to create a point at a single click rather than to draw a geometry.

```python
class Sel_Line(object):
    """Implementation for SelectFeatures_addin.selline (Tool)"""
    def __init__(self):
        self.enabled = True
        self.cursor = 3
        self.shape = "Line"
        # Can set to "Line", "Circle", or "Rectangle" for
        # interactive shape drawing.
```

This example shows the self object set to have a crosshair cursor, symbol 3 in top graphic, and allows the user to draw a line on the map.

The code shown in the following graphic also includes the handlers onCircle, onLine, and onRectangle. These handlers allow the user to draw a geometric shape on the map and return a geometric object when the task is completed. For instance, when the shape property is set to rectangle, and the user draws a rectangle on the map, an extent object is returned. Likewise, a circle returns a polygon object, and a line returns a polyline object. These objects then have their own properties, which can be accessed to determine their coordinates and other information, and can be used as a layer feature in selections.

```python
    def onCircle(self, circle_geometry):
        pass
    def onLine(self, line_geometry):
        pass
    def onRectangle(self, rectangle_geometry):
        pass
```

A variety of other handlers also report mouse clicks. The more commonly used mouse handlers are onMouseDown for pressing and holding the mouse button down, onMouseUp for releasing or letting the mouse button up, and onDblClick for double-clicking the mouse button. The mouse handlers are all shown here:

```
def onMouseDown(self, x, y, button, shift):
    pass
def onMouseDownMap(self, x, y, button, shift):
    pass
def onMouseUp(self, x, y, button, shift):
    pass
def onMouseUpMap(self, x, y, button, shift):
    pass
def onMouseMove(self, x, y, button, shift):
    pass
def onMouseMoveMap(self, x, y, button, shift):
    pass
def onDblClick(self):
    pass
```

Each handler has a function that contains code to be run when the handler is called. The self object refers to the properties of the handler, and the x- and y-values return the coordinates of the mouse position when the function is called. Note that the functions that include *Map* in their name return x- and y-values in map coordinates, whereas the other functions return x- and y-values in page coordinates.

The value for the shift property indicates which mouse button was used. The following chart gives the shift values for each mouse button. For example, if the user right-clicks on the map, the onMouseDown function returns a shift value of 2.

Button code	Mouse buttons that are pressed
1	Left button
2	Right button
3	Left and Right buttons
4	Middle button
5	Left and Middle buttons
6	Right and Middle buttons
7	All buttons

There are also handlers for keystrokes on the keyboard, as shown:

```
def onKeyDown(self, keycode, shift):
    pass
def onKeyUp(self, keycode, shift):
    pass
def deactivate(self):
    pass
```

The keycode is the ASCII code for whatever key on the keyboard is pressed, and the shift code identifies which of the Ctrl, Alt, or Shift keys were pressed simultaneously with the keystroke. The following chart shows the shift values that may be returned with one of the keystroke handlers:

Shift code	Keys that are pressed
0	No key
1	Shift key
2	Ctrl key
3	Shift + Ctrl keys
4	Alt key
5	Shift + Alt keys
6	Ctrl + Alt keys
7	Shift + Ctrl + Alt keys

Note also that there is a handler named *deactivate*. This is used to set the active tool or to see which tool is currently active. This handler can be set in code or changed when the user clicks a tool and might be used to automatically set the active tool after another action is completed. For instance, you may have the user type a map title, which then immediately activates a tool that allows the user to select features on the map.

Programming actions for the keys is as simple as adding executable code within each function. It is also recommended that you remove all the handlers that you do not intend to use from the Python script. Otherwise, all these handlers will be monitoring the user's actions continuously and activating when one of the events occurs, thus taking control of the program's process and passing it back without running any code.

With these various combinations, you can program and control events based on an almost limitless number of mouse and key actions. It is advisable, however, not to implement too many of them because it can make learning and running your application difficult.

The example design diagram at the beginning of this introduction shows that the next item to add to the Python Add-In toolbar is a combo box. Combo boxes can be used for both single typed-text entries and drop-down lists from which the user can choose a value. The creation of these elements follows the usual pattern of right-clicking the toolbar and clicking New Combo Box. As shown in the graphic, the parameters are clarified with the addition of hint text. Hint text is text that appears in the combo box when its enabled property is set to False.

Combo Box

Caption:	Select Material Type
Class Name:	SelMaterial
ID (Variable Name):	WaterAnalysis_addin.selmaterial
Tooltip:	Select the pipe material to summarize
Message:	Select from the drop-down list of material types
Hint Text:	Select Material
Help Heading:	Select the pipe material to summarize
Help Content:	Select from the drop-down list of material types

A class is added to the add-in script that contains the settings provided in addition to many properties of the self object. These parameters, included in the _init_ function, run when the combo box is first activated, as shown:

```
class Sel_Material(object):
    """Implementation for WaterAnalysis_addin.selmaterial (ComboBox)"""
    def __init__(self):
        self.items = ["item1", "item2"]
        self.editable = True
        self.enabled = True
        self.dropdownWidth = 'WWWWWW'
        self.width = 'WWWWWW'
```

The items property controls what is displayed in the combo box. In the default template, the items property is shown as a list object, and whatever is typed in the list object is displayed in a drop-down box. This property could also be set to another list item that could be derived through another means. For example, you could use the ListLayers tool from the arcpy.mapping module to find all the layers in the table of contents, and then set the self.items property equal to that list item. The drop-down box would display all the layers in the table of contents. Note that if the items property is used in the _init_ function, it is set only once. If you are making a dynamic list from a dependency as you did in tutorial 4-3, you should make the list using the onFocus function, which is described later in this introduction.

The editable property determines whether the user is allowed to type their own value or whether the user is limited to only the provided selection. If set to True, any value could be typed, even if that value does not appear in the items list. To make the combo box function a text entry box, set the enabled property to True and make the items list empty.

The enabled property is a familiar one and controls the active status of the combo box.

You can also define the width of the entry and drop-down parts of the combo box. Any letter can be used as a placeholder to designate how many characters wide either part should be. In the example, the letter *W* is used because it is the widest letter of the alphabet.

The other functions in the combo box class, shown in the graphic, handle the interaction with the user when a value is entered.

```
def onSelChange(self, selection):
    pass
def onEditChange(self, text):
    pass
def onFocus(self, focused):
    pass
def onEnter(self):
    pass
def refresh(self):
    pass
```

A combo box can be used in three ways: as a drop-down list selection, a drop-down list selection with an optional typed value, or a box in which a value can be typed. When used as a drop-down list with the editable property set to False, only the onSelChange function will work. This function returns the variable selection, which identifies which item on the list the user has selected. It can be stored in the self.selectedlayer property to be used globally, as shown:

```
def onSelChange(self, selection):
    # set selection as a global value
    self.selectedlayer = selection
```

If the combo box is used as a drop-down list selection with an optional typed value, the editable property of the combo box is set to True in the _init_ function. Users could select from the list or type their own value. To code this, you need more functions than onSelChange. The onFocus function runs when the user first clicks in the combo box. The onSelChange function runs when the user picks a value from the drop-down list. The selected value is stored in the variable selection.

If the user begins to type in the combo box, the onEditChange function runs and is followed by the onEnter function when the user clicks Enter. Note that if a value is typed, it is returned as the variable self.text in the onEditChange function or as self.value in the onEnter function. As an example, you may want to have the user select the name of an inspection officer for a report. The user can pick from the list of known inspectors or type the name of a new inspector who is filing the report. If the inspector's name is chosen from the list, the value will appear in self.selection, but if it is typed and not on the list, the value will appear as self.value. This selection method can be tricky to manage, so care must be taken to allow both types of input.

If the combo box is used as a box to type a value—that is, if the editable property is set to True and no value list is set in the items property—the user will only have the option to type a value. The onEditChange function runs when the user begins to type in the combo box, and the onEnter function stores the entered value when the user clicks Enter.

It is interesting to note that onEditChange stores the value letter by letter as it is being entered and uses a variable named *text*. You must actively store the value in the self.value property. However, the onEnter function stores the value only when Enter is pressed and stores it directly in the property self.value. The code looks like this:

```
def onEditChange(self, text):
    self.value = text
    print "The On Edit Change value is " + self.value
    pass
def onEnter(self):
    print "The On Enter value is: " + self.value
    pass
```

If you watched the user's text entry in the Python window, you would see the onEditChange function storing the value incrementally versus the onEnter function storing the value in its entirety when Enter is pressed, as shown:

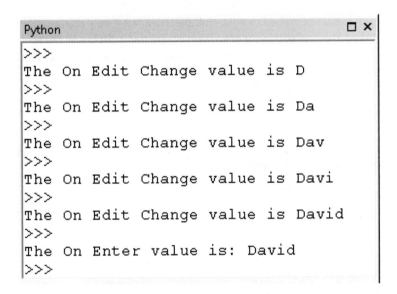

```
Python                                    □ ✕
>>>
The On Edit Change value is D
>>>
The On Edit Change value is Da
>>>
The On Edit Change value is Dav
>>>
The On Edit Change value is Davi
>>>
The On Edit Change value is David
>>>
The On Enter value is: David
>>>
```

The next item to create on the sample toolbar is a menu item. A menu resides on a toolbar and holds new buttons or additional menus. Like other items, when menus are created they require a caption and an ID, as shown:

Menu

Caption: Analysis Menu

ID (Variable Name): WaterAnalysis_addin.analysismenu

☐ Has Separator

☑ Is Shortcut Menu

The menu design screen includes two check boxes—Has Separator and Is Shortcut Menu—for characteristics that occur with menus. These check boxes are actually inactive because the behavior of the menus is controlled automatically. Every menu that appears on the main toolbar is a regular menu, requiring that you click it to display the menu choices. Then by default, every submenu is a shortcut menu, meaning that merely pausing the mouse over the main menu displays the submenu choices. Also by default, submenus do not get separators, which are vertical bar icons used to group menu items. A typical menu/submenu setup looks like this:

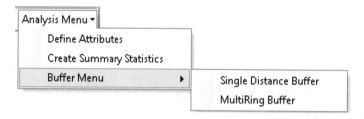

The final item to add to the sample toolbar you have been studying is a tool palette, as shown in the graphic. The palette is a special item used to hold multiple tools, but it only requires one icon on the toolbar. When clicked, the tool palette opens a drop-down menu displaying the tools you have included. After a tool is selected, the menu is compressed back to a single icon representing the selected, or active, tool.

Tool Palette

The tool palette is given the standard caption and ID, but there are two other properties to set. By default, the tools are displayed as icons only, as shown in the following graphic, and you have the option to select how many columns of icons are shown. The palette normally uses two columns, but you may select more.

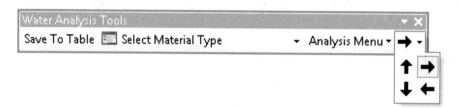

Optionally, you can set the tools to be displayed in "menu style," meaning that a single row will be presented with both the tool icon and the tool caption displayed, as shown:

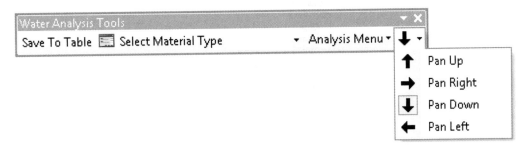

Notice that with the menu style choice, the column setting is ignored. For whichever style is chosen, however, only the selected tool remains visible on the toolbar itself.

This concludes the creation of the toolbar, and it comes close to matching the original design sketched at the beginning. As you get more practice with these toolbars and understand how each component works, you'll be able to design your toolbar's look and feel with more precision.

You may also notice that you can create extensions and menus from the Add-Ins tab on the wizard. The menus you create with the wizard can be dropped on any toolbar, not just on the custom toolbars you create. With the exception of this behavior, these add-in menus act just like any other menu, allowing you to build new buttons and submenus on them. When these menus are shared, other users can add them to any of their existing toolbars.

The extension created with the Python Add-In Wizard monitors the actions the user makes and responds to those actions, which may be to trigger an event when a new layer is added to a map document or when a specific toolbar is activated. Creating an extension is similar to other add-in items, allowing the user to set a name, class name, ID, and description, as shown:

Extension

Name:	SelectWaterLine
Class Name:	Select_Water_Line
ID (Variable Name):	WaterAnalysis_addin.selwaterline
Description:	Selects water lines in the map document

Extensions constantly monitor the user's actions, and the list in the graphic on the following page lists some of the things that an extension can react to. Some of the more common actions monitored by an extension include opening or closing a map document, starting or stopping an edit session, and adding or deleting map layers.

Methods to Implement:
- startup
- activeViewChanged
- mapsChanged
- newDocument
- openDocument
- beforeCloseDocument
- closeDocument
- beforePageIndexExtentChange
- pageIndexExtentChanged
- contentsChanged
- spatialReferenceChanged
- itemAdded
- itemDeleted
- itemReordered
- onEditorSelectionChanged
- onCurrentLayerChanged
- onCurrentTaskChanged
- onStartEditing
- onStopEditing
- onStartOperation
- beforeStopOperation
- onStopOperation
- onSaveEdits
- onChangeFeature
- onCreateFeature
- onDeleteFeature
- onUndo
- onRedo

☑ Load Automatically

The extensions you create will appear on the ArcMap Customize > Extensions menu and can be turned on or off, much like the extensions you may be more familiar with, such as the ArcGIS Network Analyst and ArcGIS Spatial Analyst extensions.

Once you have set up toolbars, menus, and extensions in the Python Add-In Wizard, you will need to add the executable code (or business logic) to each of the controls. In the working folder you identified at the start of the process is a folder named Install. Within this folder is the Python script that contains all the elements of the framework that the Python Add-In Wizard created. You can identify the classes created for each tool and add code or set parameters as necessary. You will learn more about how to do this in this chapter's tutorials.

When you start writing the code for your add-in, there are special functions you can use for dialog and pop-up message boxes. These functions are included in the Python add-ins module (pythonaddins), which is imported in the same manner as the ArcPy module (arcpy), as shown:

```
# Import ArcPy and Python Add-ins modules
import arcpy
import pythonaddins
```

Two of these functions allow the user to interact with data files using either an input dialog box to select features as input or an output dialog box to save the output of your script. The first, the OpenDialog() function, accepts a dialog box title to be displayed along the top of the dialog box, a property to allow multiple file inputs, the path for the input data, and a caption for the dialog box button. In the following graphics, note the code and the dialog box it creates to allow for multiple input layers to be selected. The dialog box appears when the user clicks the button and activates the onClick function, as shown:

```python
class ButtonClass1(object):
    """Implementation for ExampleToolbar_addin.mybutton1 (Button)"""
    def __init__(self):
        self.enabled = True
        self.checked = False
    def onClick(self):
        # OpenDialog({title}, {multiple_selection}, {starting_location}, {button_caption})
        pythonaddins.OpenDialog("Select Layers", True, r"C:\EsriPress\GISTPython\Data\Sample Data.gdb",\
        "Process")
```

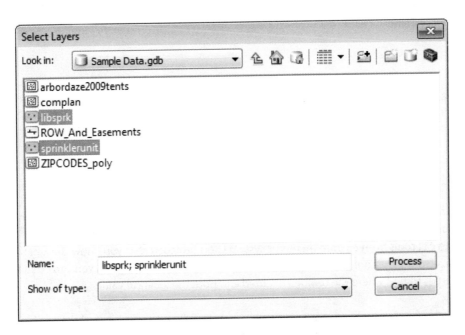

The second function allowing interaction with data files is the SaveDialog() function, which opens a similar navigation box to that of the OpenDialog() function. It can be activated from within an add-in button using the onClick function and provides a dialog box title, a save location, and a default name for the feature. Note the code shown here and the dialog box it creates, as shown on the next page:

```python
class ButtonClass1(object):
    """Implementation for ExampleToolbar_addin.mybutton1 (Button)"""
    def __init__(self):
        self.enabled = True
        self.checked = False
    def onClick(self):
        # SaveDialog({title}, {name_text}, {starting_location})
        pythonaddins.SaveDialog("Save New Layers", "Results of Process", \
        r"C:\EsriPress\GISTPython\MyExercises\MyAnswers.gdb",)
```

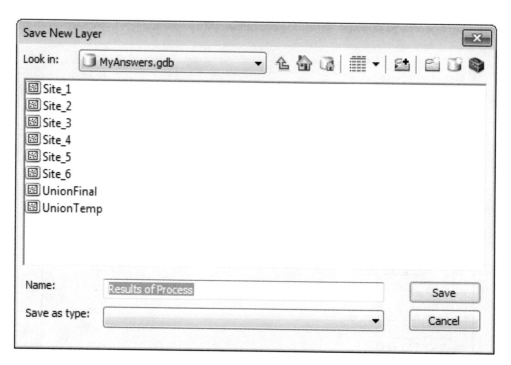

One of the more interesting functions in the Python Add-ins module, the GPToolDialog() function, calls any geoprocessing tool from an add-in button. The code basically names the tool and the toolbox it comes from, and the tool's dialog box is displayed when the button is clicked. This can be used to call tools from custom menus or system tools provided that you know the location of the toolbox storing the tool you wish to use. To find the toolbox system path, you can open the toolbox properties to find where the toolbox is stored and note the location value.

A sample of code to call a custom tool is shown in the following graphic, and the resulting dialog box is exactly the same as if the tool were run from the Catalog window.

```
class ButtonClass1(object):
    """Implementation for ExampleToolbar_addin.mybutton1 (Button)"""
    def __init__(self):
        self.enabled = True
        self.checked = False
    def onClick(self):
        # GPToolDialog(toolbox, tool_name)
        pythonaddins.GPToolDialog(r"C:\EsriPress\GISTPython\MyExercises\Custom Python Tools.tbx",\
        "Count Buildings")
```

The Python Add-Ins module also has its own type of message window. This window allows you to display a notice to the user with a pop-up dialog box as a result of some action in the script. The message box parameters include the message to display, a title for the message box, and, more importantly, the type of message box to display. There are seven different types of message boxes, which can be used to prompt the user to continue with the process, to cancel, to retry, and other choices. The options are shown in the chart:

mb_type code	Message Box Type
0	OK only
1	OK/Cancel
2	Abort/Retry/Ignore
3	Yes/No/Cancel
4	Yes/No
5	Retry/Cancel
6	Cancel/Try Again/Continue

The sample code shown in the following graphic checks to see how many features are selected and warns the user that an amount over 500 may take an excessive amount of time to process. The OK/Cancel type of dialog box used allows the user to either continue the process if they are willing to wait or to cancel the process.

```
# Count the features, and warn the user if the number is over 500
for lyr in lyrList:
    if lyr.isFeatureLayer:
        count = arcpy.GetCount_management(lyr).getOutput(0)
        if count > 500:
            # MessageBox(message, title, {mb_type})
            pythonaddins.MessageBox("More than 500 features are selected. This may take a long time to process.",\
                            "Large Selection Warning",1)
```

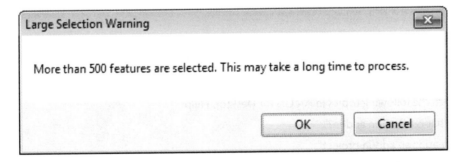

The final function in this module allows the script to access the currently selected layer from the map's table of contents or the currently active data frame in your map document. This function may be helpful if the user has made this selection or if it is already known which layer or data frame is selected. The script should create a map document object and then call the GetSelectedTOCLayerOrDataFrame() function to get the active data frame. To get the currently selected layer, a data frame object must be defined before calling this function. The sample code in the graphic on the following page shows how to create a data frame object using the active data frame.

```
def onClick(self):
    # Get the current document
    mxd = arcpy.mapping.MapDocument("CURRENT")
    # Get the currently active data frame
    # GetSelectedTOCLayerOrDataFrame()
    activeDF = pythonaddins.GetSelectedTOCLayerOrDataFrame()
```

All the work completed so far creates only the toolbar structure and executable code. The add-in must still be compiled and an .addin file created, which is used to install the add-in in ArcGIS. The folder storing the add-in components has a prewritten script named makeaddin.py, and running this script compiles and creates the .addin file. Once it is created, this single file is all that is needed to implement your project or share your toolbar with others.

Tutorial 5-1 Creating a Python add-in application

Creating a Python add-in is a great way for users to design and implement a complex GIS application. Add-in applications are compiled into a single file that is easy to share and can contain all the tools necessary for a project.

Learning objectives

- Designing a user interface
- Creating a Python add-in
- Creating a button

Preparation

Research the following topics in ArcGIS for Desktop Help:

- "What is a Python add-in?"
- "Creating an add-in project"
- "Creating a Python add-in button"

Introduction

The process of creating a Python add-in should not be taken lightly. Add-ins you create are fully developed applications that can contain their own menus and toolbars, with various components, and sets of executable code to accomplish various geoprocessing and data analysis tasks. The design is more than just writing pseudo code; it requires that you fully design a user interface to contain all the tools that your add-in will control. This design can then be used to create the framework and code for your add-in. It is important to note that changing the framework of an add-in is almost like starting over and should be avoided. Take the time to fully design your interface before you start.

Scenario

Realizing that there is no easy way to see how many features are in a layer before starting a long analysis process, you decide to make a tool for this purpose. Because you want to be able to share this tool easily with others, you will make it a Python add-in application. The process will be to iterate through the layers in the current map document and display a feature count in a pop-up window.

Data

There is no restriction on what type of data to use because the add-in, by nature, will look at every layer in the table of contents. Any map document will do, but a reasonably sized map document is provided for testing. The toolbar design will have a single button with the title "Get Feature Count."

SCRIPTING TECHNIQUES

The special introduction to Python add-ins in this chapter shows how to create the template folder and files using the wizard application. For this tutorial, identify the different classes the tool contains. The first is the Toolbox Class, which holds all the information about the toolbar.

The next class is the one created to hold the information about the button you are creating. Within this class are many functions, each pertaining to an action the user can take by clicking the button. The two functions of importance here are the _init_ function, which initializes the button, and the onClick function, which contains the code to run when the button is clicked.

Use the Python Add-In Wizard

1. Unzip the addin_assistant.zip file in the GISTPython\Python Add-Ins folder. Move to the bin folder in the unzipped files, right-click addin_assistant.exe, and click Send to Desktop to create a shortcut. Note: the process for creating a desktop shortcut may differ depending on your operating system. Rename the shortcut Python Add-In Wizard.

2. Double-click the new desktop icon for the Python Add-In Wizard to start the process of creating the Python add-in.

3. Navigate to your MyExercises folder, and click the Make New Folder button at the bottom of the dialog box. Enter the name of the new folder as Feature Count and click OK, as shown:

With the working folder set, move on to setting the project properties.

4. Set the product to ArcMap, and fill in the other values as shown in the graphic. For the image, navigate to the Data folder, and select the City of Oleander logo (OleaderTX.png). When you have the values entered, click Save.

5. Click the Add-In Contents tab. Create a new toolbar by right-clicking Toolbars and clicking New Toolbar. Fill in the caption and ID as shown in the next graphic. Note that the entire default string for the ID was removed, and a new name was given. When you move to the Description entry box, the ID prefix is automatically added. Leave the Show Initially box checked to show the toolbar by default when ArcMap is started, as shown:

Toolbar

Caption:	Count the features
ID (Variable Name):	countfeatures
	☑ Show Initially

6. **Right-click the new toolbar and click New Button. Fill in the caption and other properties as shown. Leave the image blank.**

Button

Caption:	Count Features
Class Name:	Count_Features
ID (Variable Name):	FeatureCount_addin.countbutton
Tooltip:	Get a count of features
Message:	Count all features for layers in the TOC
Help Heading:	Get a count of features
Help Content:	Count all features for layers in the TOC
Image for control:	

One toolbar with a single button on it is now created. Next, save the toolbar, and add the code for the button.

7. **Click Save and click the Open Folder button. Navigate to the Install folder, and open FeatureCount_addin.py in your IDE, as shown:**

```
import arcpy
import pythonaddins

class Count_Features(object):
    """Implementation for FeatureCount_addin.countbutton (Button)"""
    def __init__(self):
        self.enabled = True
        self.checked = False
    def onClick(self):
        pass
```

You can see that the script brings in the ArcPy and Python add-ins modules. A class named Count_Features has been created (per the class name you set for the button), and a couple of default properties have been set. All you have to do to make this button perform a task is add code to the onClick function. The code has been provided, but feel free to tackle the code portion of this tutorial on your own for an extra challenge. Here's the pseudo code for the button's task:

```
# Create an object to store the current map document.
# Create an object to store the current data frame.
# Create a list object of all the layers in the current map document.
# Create an empty variable to hold the final display message.
# Using a for statement, iterate through the list of layers.
# Get a count of the layer's features.
# Generate a text message reporting the count and the layer name.
# Note that this text message will be appended to the message variable
#    and should include a carriage return at the end of the message.
# Display the message in a Python window in ArcMap.
# Display the final results message in a pop-up message box.
```

8. Open the text file Tutorial 5-1 in the Data folder to view the code for the button, as shown:

```python
"""Code to count features in a layer:"""

        # Get the current document.
        mxd = arcpy.mapping.MapDocument("CURRENT")
        # Get the current data frame.
        df = arcpy.mapping.ListDataFrames(mxd)[0]
        # Make a list object of all the layers in the TOC.
        lyrList = arcpy.mapping.ListLayers(mxd, "", df)
        # Initialize the message variable as empty.
        msg = ""
        # Iterate through the list of layers, and perform the count.
        # Add the count to the message variable - note the line return.
        for lyr in lyrList:
            if lyr.isFeatureLayer:
                count = arcpy.GetCount_management(lyr).getOutput(0)
                msg = msg + "Layer " + lyr.name + " contains " + str(count) + " features.\n"
                print "Layer " + lyr.name + " contains " + str(count) + " features."
        # Message is printed to the Python box but also displayed in an add-in pop-up message.
        pythonaddins.MessageBox(msg, "Layer feature counts in: " + df.name, 0)
```

9. If you are not writing your own code, copy the code from the text file, and paste the code to the FeatureCount_addin.py file inside the onClick function, as shown. Make sure to correct any indentation errors, and then save and close the Python script.

```python
import arcpy
import pythonaddins

class Count_Features(object):
    """Implementation for FeatureCount_addin.countbutton (Button)"""
    def __init__(self):
        self.enabled = True
        self.checked = False
    def onClick(self):
        # Get the current document.
        mxd = arcpy.mapping.MapDocument("CURRENT")
        # Get the current data frame.
        df = arcpy.mapping.ListDataFrames(mxd)[0]
        # Make a list object of all the layers in the TOC.
        lyrList = arcpy.mapping.ListLayers(mxd, "", df)
        # Initialize the message variable as empty.
        msg = ""
        # Iterate through the list of layers, and perform the count.
        # Add the count to the message variable - note the line return.
        for lyr in lyrList:
            if lyr.isFeatureLayer:
                count = arcpy.GetCount_management(lyr).getOutput(0)
                msg = msg + "Layer " + lyr.name + " contains " + str(count) + " features.\n"
                print "Layer " + lyr.name + " contains " + str(count) + " features."
        # Message is printed to the Python box but also displayed in an add-in pop-up message.
        pythonaddins.MessageBox(msg, "Layer feature counts in: " + df.name, 0)
```

10. Close the text file and the Python Add-In Wizard. In Windows Explorer, navigate to the Feature Count folder. Run the Python script makeaddin.py to create the Feature Count add-in, and note the file that is created, as shown:

Name	Date modified	Type
Images	1/20/2014 8:33 PM	File folder
Install	1/20/2014 8:33 PM	File folder
config	1/20/2014 8:36 PM	XML Document
Feature Count	1/20/2014 8:38 PM	Esri AddIn File
makeaddin	1/20/2014 8:33 PM	Python File
README	1/20/2014 8:33 PM	Text Document

11. Double-click Feature Count.esriaddin and click Install Add-In, as shown. Click OK when notified of a successful installation.

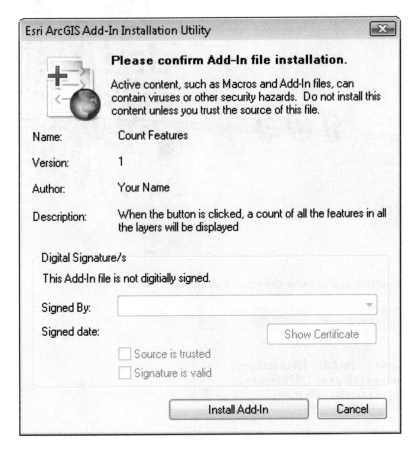

The add-in is now installed and ready to test.

12. Open the map document Tutorial 5-1, as shown. The new toolbar should be loaded automatically, but if it is not, go to Customize > Toolbars to add it.

13. Click the Count button, and observe the results, as shown:

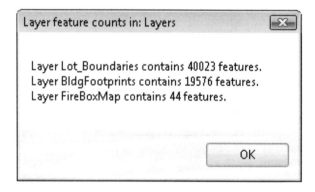

Use your new Count tool to get feature counts in other map documents.

Exercise 5-1

The feature count is working great and is a big success, but there is another request for a summary from the table of contents. Create a new toolbar with a button that gives a summary of how many of each of the major feature type layers there are in the table of contents. For example, the message box would report how many of each polygon, line, and point layers were found in the table of contents. As a bonus, make it flexible enough to report all the layer types.

Tutorial 5-1 review

This tutorial demonstrates the basic template for creating Python add-ins. After sketching a design, you can create the toolbars, buttons, and menus, and a Python script is created.

Each item you define in the creation wizard has its own class in the add-in's script. Different types of features have different functions to handle their actions, and this class was quite simple with an onClick function. Within that function, you can add code to perform a task—something as simple as counting the features in a feature class or as complex as your coding skills allow.

Study questions

1. Will the add-in you wrote for tutorial 5-1 work with any map document? Why or why not?
2. What file(s) would you need to share with other users who wanted this add-in?
3. When does the _init_ function run, and what types of tasks might you have it perform?

Tutorial 5-2 Using buttons and combo boxes

Python add-ins allow you to design a complete menu bar for an application you create. An add-in can consolidate custom and system tools in one place and contain a variety of buttons and combo boxes.

Learning objectives

- Designing a custom menu bar
- Designing a menu combo box
- Modifying a user interface

Preparation

Research the following topics in ArcGIS for Desktop Help:

- "Creating an add-in toolbar"
- "Creating a Python add-in combo box"

Introduction

In tutorial 5-1, you learned that the Python Add-In Wizard lets you design your toolbars and menus, with all the associated components, and then write a template script with all the classes and handlers. This script can be edited, and your own code can be added to the various functions to control the application. One major concern is that if you then go back and change your project design, a new template file will be generated without all the extra code you have written. A backup of the old file is automatically created, and you have to copy all your code from there to the new file. The inconvenience of doing this highlights the importance of designing your project in full the first time.

The toolbar can hold a variety of items, as shown in the special introduction to this chapter, and in this tutorial, you will be working with a combo box. A combo box can be used as a drop-down list or as a text-entry box, depending on the status of the edit property and the list you create. Either way, the result is a value that you can then pass to other functions in the application.

You may also want to use one item's action to control another's. For instance, leave a button disabled until an action is taken by another item. You may have a combo box accept a buffer distance, and the button to start the buffering process would be disabled until an appropriate value is selected.

Scenario

The users are happy with the toolbar from tutorial 5-1, except that if there are many layers in the table of contents, the drop-down box will produce a rather long list of layers. Then the script takes time to read the list and find the layer of interest. The application would be more functional if the user could select a single layer and get a feature count for just that layer. Add a combo box that provides a list of layers from the table of contents so that the user can select from this list. Then a count of the features can be retrieved from just the selected layer. In addition, add a button that lets the user browse to any workspace, select a feature class, and report back the number of features without having to add the feature class to the current map document.

Data

A completed copy of the application to count the features is provided in the Python Add-Ins folder named SelectNCount. Modify this application to include a combo box and another button.

SCRIPTING TECHNIQUES

The first new thing you will work with is the Add-In Manager in ArcMap, which is an interface that allows you to see what add-ins are currently loaded and to remove any that are no longer necessary. Included in the Add-In Manager, by default, is the ArcGIS Online add-in. You should leave that add-in intact.

You will also see how to access an existing application with the Python Add-In Wizard. At the wizard start-up, point it to the folder of an existing add-in, and the wizard will reload all the items that were designed earlier. Some of the changes, such as tooltips and descriptions, can be altered and saved back to the original files. But if you change any of the design items, a new template is generated. These design items include the caption and class properties of any item and the addition or deletion of any item, such as menus, toolbars, and any of the items that these can contain.

If the core design is altered and a new template is generated, a backup of the original file is automatically saved as the original name with _1 appended to it. This number increases incrementally if additional changes are made. You would then need to copy and paste the custom code that you added to the template from the old file to the new file. Depending on the purpose of the change, you might also have to alter the code to accommodate the new items.

As you saw before, each item creates a new class in the Python script. It may be necessary to pass values from one function to another, and that is done with the ID (variable name) you provided each item when it was created. These IDs appear in a comment line for each class as a reminder. Properties for other functions can also be controlled. For instance, to disable the button in a function with the button ID of startbuffering, you would set the value "self.enabled = False" in the button's function, and then add the code "startbuffering.enabled = True" to the other, controlling function.

The combo box you add will include code to get a list of layers from the data frame, so first create a map document object and use it in the creation of a data frame object. You can then use these objects to create a list object of the included layers. You have created these objects in other scripts, and you can reference them for the syntax of these commands.

The new button that you add, however, will use a new technique for getting user input. It will open a dialog box using the OpenDialog() function from the Python add-ins module. The introduction to this chapter shows how to format this function correctly, and it will allow the user to browse to a workspace to select a feature class.

Modify an existing Python add-in

1. Open the map document Tutorial 5-2. On the main menu, click Customize > Add-In Manager, as shown. You may see several add-ins loaded. One by one, select and delete these add-ins, leaving only ArcGIS Online. Then close the Add-In Manager and exit ArcMap.

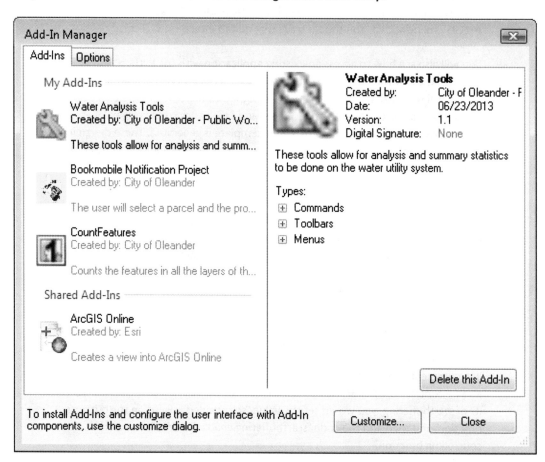

The editing and troubleshooting process needs to take place with ArcMap closed. Then as the changes are made, the add-in will be recompiled, and you will start ArcMap to test the latest changes. It is necessary to edit with ArcMap closed and open it when you are finished because the add-in reinitializes and loads the new changes only when ArcMap first starts up.

2. Go to the folder where you installed the sample data, and copy the folder SelectNCount from the Python Add-Ins folder to your MyExercises folder.

3. Start the Python Add-In Wizard, and point it to the SelectNCount project in your MyExercises folder. Note that it has loaded the properties for the existing add-in, as shown:

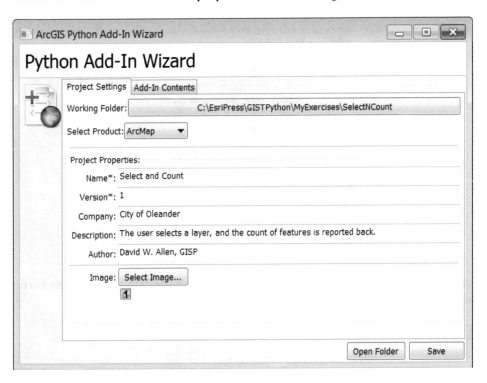

4. Click the Add-In Contents tab, and add a new combo box to the existing toolbar.

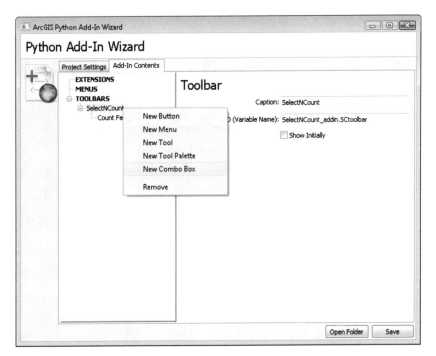

5.   Set the characteristics for the combo box as shown:

Combo Box

Caption:	Select Layer
Class Name:	Select_Layer
ID (Variable Name):	SelectNCount_addin.selectlayer
Tooltip:	Select Layer
Message:	Select a layer from the list to count
Hint Text:	Select Layer
Help Heading:	Select Layer
Help Content:	Select a layer from the list to count

6.   Without clicking Save, right-click the SelectNCount toolbar item and click New Button. Populate the button characteristics as shown:

Button

Caption:	Select Feature Class
Class Name:	Select_Feature_Class
ID (Variable Name):	SelectNCount_addin.selectfeatclass
Tooltip:	Select Feature Class
Message:	Browse to a workspace and select a feature c
Help Heading:	Select Feature Class
Help Content:	vorkspace and select a feature class to count
Image for control:	

7. Click Save. A warning box appears commenting on the creation of a new script for the application. Note the warning shown and click OK.

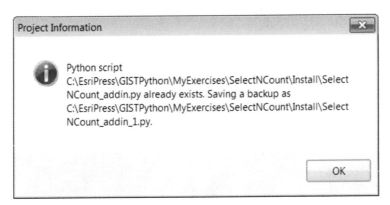

The warning lets you know that a new template file has been created. You will have to copy your code from the old file to the new one.

8. On the wizard dialog box, click Open Folder, and open the Install folder. You will see both script files. Open the SelectNCount_addin.py file in your IDE, and notice that none of your code was copied over to the new file, as shown:

```python
import arcpy
import pythonaddins

class Count_Features(object):
    """Implementation for SelectNCount_addin.countfeatures (Button)"""
    def __init__(self):
        self.enabled = True
        self.checked = False
    def onClick(self):
        pass

class Select_Feature_Class(object):
    """Implementation for SelectNCount_addin.selectfeatclass (Button)"""
    def __init__(self):
        self.enabled = True
        self.checked = False
    def onClick(self):
        pass

class Select_Layer(object):
    """Implementation for SelectNCount_addin.selectlayer (ComboBox)"""
    def __init__(self):
        self.items = ["item1", "item2"]
        self.editable = True
        self.enabled = True
        self.dropdownWidth = 'WWWWWW'
        self.width = 'WWWWWW'
    def onSelChange(self, selection):
        pass
    def onEditChange(self, text):
        pass
    def onFocus(self, focused):
        pass
    def onEnter(self):
        pass
    def refresh(self):
        pass
```

Notice that new classes for the combo box and second button were added. These are the basic item templates that you update with code to perform the tasks you want. Write the code for the combo box to select layers from the table of contents, and then write the code for the button to select a feature class from any workspace, as shown:

```
class Select_Feature_Classes(object):
    """Implementation for SelectNCount_addin.selectfeatclass (Button)"""
    def __init__(self):
        self.enabled = True
        self.checked = False
    def onClick(self):
        pass

class Select_Layer(object):
    """Implementation for SelectNCount_addin.selectlayer (ComboBox)"""
    def __init__(self):
        self.items = ["item1", "item2"]
        self.editable = True
        self.enabled = True
        self.dropdownWidth = 'WWWWWW'
        self.width = 'WWWWWW'
    def onSelChange(self, selection):
        pass
    def onEditChange(self, text):
        pass
    def onFocus(self, focused):
        pass
    def onEnter(self):
        pass
    def refresh(self):
        pass
```

In the previous version of this script, a list of layers was generated when the button was clicked. In this version, a list of layers is created when the user first clicks in the combo box. This is controlled by the onFocus function. By generating the list in the combo box, the onFocus function will update the table of contents with any changes the user makes.

The onFocus function has an object named focused that is true if the cursor is currently positioned in the combo box and false if it is elsewhere. You can use that value for the trigger to create the selection list. The list is created in the same way as in the original script. Then the list of values for the combo box is stored in the self.items object. The trick here is that the list wants the layer names, not the layer objects that the list object is storing. To get the layer names, you can use a for statement to iterate through the list and add the names to self.items. There is an additional piece of logic in the code, shown in the graphic for step 9, that checks to make sure the layer contains features so that the count will not be done on layers such as annotation or dimension.

9. Write the code for the onFocus function in the Select_Layer class to populate the list of values. Try this on your own first, and then compare your code to the code shown:

```
def onFocus(self, focused):
    #Populate the items list with the available layers
    if focused:
        thisMap = arcpy.mapping.MapDocument("CURRENT")
        layers = arcpy.mapping.ListLayers(thisMap)
        self.items=[]
        for lyr in layers:
            if lyr.isFeatureLayer:
                self.items.append(lyr.name)
```

The function onSelChange is called when the user makes a choice from the value list. The function has a parameter named selection that holds the value, which can then be saved to the self object as self.selectedlayer. This object is then used to pass the value to the button where the count process occurs.

In addition, there is a line of code to set the button's enable property to True. In other words, the button is disabled until a selection is made.

10. Find the onSelChange function in the Select_Layer class. Add the code to save the selection and enable the button, as shown:

```
def onSelChange(self, selection):
    # Set the global value (becomes selectlayer.selectedlayer)
    self.selectedlayer = selection
    # Enable the button for counting
    countfeatures.enabled = True
```

The final change for the combo box is to make sure the items list is blank when it initializes. The only change is to remove the sample values from the self.items list object.

11. Change the self.items list object to empty, as shown:

```
class Select_Layer(object):
    """Implementation for SelectNCount_addin.selectlayer (ComboBox)"""
    def __init__(self):
        self.items = [] # The list will start empty
        self.editable = True
        self.enabled = True
        self.dropdownWidth = 'WWWWWW'
        self.width = 'WWWWWW'
```

This combo box enables the Count button when a valid entry is selected, so you must alter the code for this button to be disabled when it initializes. Use a pop-up message box to display the count to the user.

12. Find the _init_ function of the Count_Features class, and change the enabled property of the button to False, as shown:

```
class Count_Features(object):
    """Implementation for SelectNCount_addin.countfeatures (Button)"""
    def __init__(self):
        self.enabled = False
        self.checked = False
```

The main processes of the Count button occur in the onClick function. The function brings in the selected value from the Select_Layer class, using the ID selectlayer, and uses that value for the feature count process. The results can be displayed in a pop-up message box.

13. Find the onClick function of the Count_Features class, and remove the code word *pass*. Replace the code word with code to transfer the value to the onClick function, and show the results in a Python add-ins message box. Check your code against the code shown:

```
def onClick(self):
    infeatures = selectlayer.selectedlayer
    count = arcpy.GetCount_management(infeatures).getOutput(0)
    msg = "Layer " + infeatures + " contains " + str(count) + " features."
    print "Layer " + infeatures + " contains " + str(count) + " features."
    # Message is printed to the Python box but also displayed in an add-in pop-up message
    pythonaddins.MessageBox(msg, "Feature Count", 0)
```

If you like, you can save the script, compile the add-in, and test with the map document Tutorial 5-2. At this point, the Select Feature Class button will not have any action associated with it, but the combo box and count button should work.

Configuring the dialog box to accept user input for counting uses the pythonaddins.OpenDialog() tool described in the special introduction to this chapter. The syntax is to make this tool equal to a variable, which will become a feature class object containing the parameters of the selected feature class.

14. Find the onClick function in the Select_Feature_Class class. Add a variable named inFeatures, and make the variable equal to the OpenDialog tool with the appropriate parameters. The starting workspace should be the Sample Data geodatabase, as shown:

```
class Select_Feature_Class(object):
    """Implementation for SelectNCount_addin.selectfeatclass (Button)"""
    def __init__(self):
        self.enabled = True
        self.checked = False
    def onClick(self):
        # OpenDialog({title}, {multiple_selection}, {starting_location}, {button_caption})
        inFeatures = pythonaddins.OpenDialog("Select Feature Clas",False,\
                r"C:\EsriPress\GISTPython\Data\Sample Data.gdb","Select")
```

The process to get the count and display it in a pop-up box is exactly the same as that used for the Count button.

15. In the Count Features class, find the five lines of code that perform the counting, starting with the line that includes the GetCount() function down to the line that includes the MessageBox() function. Copy these lines of code from this class and paste them to the onClick function of the Select_Feature_Class class, as shown:

```
def onClick(self):
    # OpenDialog({title}, {multiple_selection}, {starting_location}, {button_caption})
    inFeatures = pythonaddins.OpenDialog("Select Feature Class",False,\
                r"C:\EsriPress\GISTPython\Data\Sample Data.gdb","Select")

    count = arcpy.GetCount_management(inFeatures).getOutput(0)
    msg = "Layer" + inFeatures + " contains " + str(count) + " features."
    print "Layer " + inFeatures + " contains " + str(count) + " features."
    # Message is printed to the Python box but also displayed in an add-in pop-up message
    pythonaddins.MessageBox(msg, "Feature Count", 0)
```

16. Save the script, and close your IDE. Go back to the SelectNCount folder—click Open Folder in the wizard dialog box if necessary. Run the makeaddin.py script, and then double-click the SelectNCount.esriaddin file to install the add-in, as shown:

Name	Date modified	Type	Size
Images	6/25/2013 9:10 PM	File Folder	
Install	6/25/2013 9:17 PM	File Folder	
config.xml	6/25/2013 9:16 PM	XML Document	2 KB
makeaddin.py	6/25/2013 9:08 PM	PY File	2 KB
README.txt	6/25/2013 9:08 PM	Text Document	1 KB
SelectNCount.esriaddin	6/25/2013 11:42 PM	Esri AddIn File	6 KB

17. Open the map document Tutorial 5-2. If the toolbar was not installed automatically, add it from the Customize menu. Test the Select Layer combo box drop-down list by selecting a layer and then clicking the Count Features button to get the count, as shown:

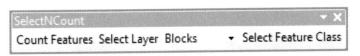

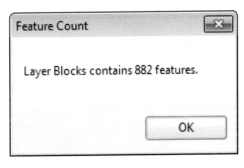

Notice that the Count Features button is disabled until a selection is made in the combo box.

18. Click the Select Feature Class button to test it. Select one of the layers in the default location, as shown in the graphic, and then click the Count Features button.

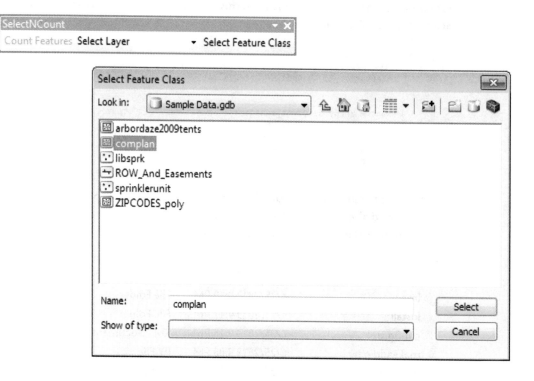

You may want to go back and adjust the widths of the combo box and drop-down list so that the entire layer names will show. Check the special introduction to this chapter for information on how to do this. You may also want to try navigating to a different workspace using the Select Feature Class button and see how the tool responds.

Exercise 5-2

Add a button to the application you built in exercise 5-1 to allow for a similar layer selection from a browsing-style dialog box. The user will click the first button, which will open a dialog box and allow the user to browse to a workspace. Clicking the Count Features button will produce a count of features for all the feature classes in the selected workspace. As in exercise 5-1, display the results grouped by feature class type (point, line, or polygon). As a bonus, use the describe object .dataType method to validate that the user's selection is a workspace because this application will operate only on workspaces. Also, use the .children method to list any child items or feature datasets.

Tutorial 5-2 review

As you have learned, making changes to the layout of an add-in can be problematic. The add-in wizard allows you to add additional items to a toolbar you have already created, but then the wizard writes a completely new template file. The problem is that it does not copy any of your code to the new file; that task is left for the programmer to do manually. The example used had one simple button to transfer, but imagine doing this for a toolbar that has six to 10 controls. Learning this labor-intensive process serves to emphasize the importance of designing your interface completely when you write your pseudo code so that you do not find yourself facing a similar situation.

This add-in includes a combo box. The interesting technique here is to populate the combo box with a list of feature classes from which the user can select. Then once an item is selected from the list, the value can be passed to other functions and used in your code.

The add-in also uses a user input dialog box from the Python add-ins module. This project uses the dialog box to accept input, but there is also a tool that will accept and create an output file. The results are displayed in a pop-up message box, which is useful for displaying a message, but with the correct type set, message boxes can also be used for error handling. Any of the types that contain a Cancel button will cause the script to end, but only by the choice of the user, which is not the same as detecting an issue with a script and automatically ending it as you might with a try-except code block.

Study questions

1. What other lists might you put in a combo box?
2. Why was altering the template of the add-in such a problem?
3. How was the onClick function able to get the value of the selected feature from the other class?

Tutorial 5-3 Using tools to interact with the map

Python add-ins have the unique capacity of offering the user direct interaction with the map through a Python tool. This tool allows the program to control all phases of the user's experience with the application.

Learning objectives

- Designing an add-in interface
- Using Feature Selection tools

Preparation

Research the following topics in ArcGIS for Desktop Help:

- "Creating an add-in toolbar"
- "Creating an add-in tool palette"
- "Creating a Python add-in tool"

Introduction

Add-ins allow you to program and add a variety of items to a custom toolbar and share the resulting application with others. Add-ins do one thing that none of the other ArcGIS customization methods do—they allow the user to interact with features on the map. In all the other scripts, script tools, and Python toolboxes that you have written in this book, it was always a requirement to have the user make their selections *before* running the script. That is no longer the case when you add a tool to an add-in. The tool can return its location on the map, and that location can be used for selections and other location-based tasks.

Scenario

The application you wrote for the real estate agents in tutorial 3-2 was a big hit. If you recall, this application allowed the agents to enter the tax ID of the subject tract, a map title, and a subtitle and then select between four different output maps. In this tutorial, you will work with three of these maps: a water map, a storm water map, and a sewer map. By taking this application and making it a Python add-in, you will be able to add the ability for the agents to interactively select the subject tract on the map by clicking the desired tract. The toolbar will look like this sketch:

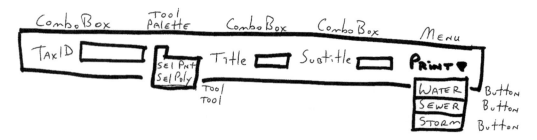

Create a toolbar with the following elements:

- Combo box—to accept the typed tax ID
- Tool palette—to hold the two selection tools
- Tool (in a tool palette)—to allow the user to select a single feature with one click
- Tool (in a tool palette)—to allow the user to select multiple features using a rectangle
- Combo box—to accept the map title
- Combo box—to accept the map subtitle
- Menu—to hold the print buttons
- Button—on the menu to print the water map exhibit to a PDF
- Button—on the menu to print the sewer map exhibit to a PDF
- Button—on the menu to print the storm water exhibit to a PDF

As a bonus, if you are comfortable working with the config.xml file, add a Clear Selected Features system button. It would also be nice to disable the combo boxes for the title and subtitle until a valid selection is made.

This list of toolbar elements should give you an idea of what needs to be done in the Python Add-In Wizard to create the complete template and of where to add the code.

Data

A map document is provided for you to test, but you must develop all the code from scratch. You can use the sample add-in mentioned in the following "Scripting techniques" section to get code for the selection buttons and also reference your code from tutorial 3-2.

You should write pseudo code for each of the toolbar items describing what operation the tool will perform and how it may interact with the enabling or disabling of other items on the toolbar.

SCRIPTING TECHNIQUES

The ability to select features on the map using tools from your add-in is an element that you will want to incorporate in your scripts as much as possible to make your tools more user-friendly. There are a few obstacles to overcome, but a sample add-in is provided so that you can see how to overcome them.

The way to perform an interactive selection on the data in a map document is to use an add-in tool to capture the screen location, and then build objects from that location's coordinates that can be used in a SelectByLocation process. If you look at the default template created with new add-in tools, you will see many different functions that are called when the user clicks the mouse, as shown in the special introduction to this chapter. Two of the functions are called when the mouse is clicked down, and both return x,y coordinates: the onMouseDown function and the onMouseDownMap function. However, these functions provide different information. The onMouseDown function provides coordinates in screen pixels measured from the upper-left corner of your computer screen. This is of no use when working with the map. The other choice is the onMouseDownMap function, which provides map-based coordinates. The problem is that the coordinates are dependent on the view you are currently using. If you are in data view, the returned coordinates are in the projected map units of the current map document. These coordinates can be used directly and will create objects that overlay your data. But when you are in layout view, the returned coordinates are in page units and will not overlay the map data. The page coordinates can be converted to projected map units with a little code, though, and then used to make selections on the map.

Other tools that capture a geometric shape drawn by the user, including a circle, a line, or a rectangle, are called *click event handlers*. When the user draws a geometric shape on the map, the click event handlers store the properties of the shape drawn. The coordinates for the geometric shape can then be used to help select features on the map by using the coordinates to create temporary layer objects and overlaying them on other layers. The click event handlers for this action are onCircle, onLine, and onRectangle, or if the click event handler is set to NONE, a single point is used for the selection.

To see the point and rectangle click event handlers in action, go to the GISTPython\ Python Add-Ins folder and double-click the add-in file SelectFeatures. Then open the map document Parcels from the Maps folder. Add the toolbar Select Features, if necessary. These two tools will let you select features from the Parcels layer with either a single mouse click, using the onMouseDownMap function as a click event handler, or by dragging a box around them, using the onRectangle function as a click event handler, similar to the way features are selected with the system selection tools. Note that you can use these tools in both data and layout views because the script has code added to convert the coordinates to projected map coordinates before making the selection.

The code for selections is provided in the Python Add-Ins\SelectFeatures\ Install folder. You can open the SelectFeatures_addin.py file and see how the conversion process takes place. The conversion code is in a function named convertCoords at the top of the script. The function can be called from elsewhere in the script and accepts a pair of coordinates in page units, converts the page units to projected map units, and returns the new values back to the line of code that called the function. Examine the code shown in the graphic, and note how the position relative to the page is converted into projected map units, using the map scale and data frame coordinates. This function can be added to all your projects to perform the coordinate conversions if features are selected in page units.

```
def convertCoords(x,y):
    thisMap = arcpy.mapping.MapDocument("CURRENT")
    dataFrame = arcpy.mapping.ListDataFrames(thisMap)[0]
    pageX = x # The function accepts the X coordinate sent from the calling function.
    pageY = y # The function accepts the Y coordinate sent from the calling function.

    """Convert page coordinates to projected coordinates"""

    # Get the data frame dimensions in page units.
    df_page_w = dataFrame.elementWidth
    df_page_h = dataFrame.elementHeight
    df_page_x_min = dataFrame.elementPositionX
    df_page_y_min = dataFrame.elementPositionY
    df_page_x_max = df_page_w + df_page_x_min
    df_page_y_max = df_page_h + df_page_y_min

    # Get the data frame projected coordinates.
    df_min_x = dataFrame.extent.XMin
    df_min_y = dataFrame.extent.YMin
    df_max_x = dataFrame.extent.XMax
    df_max_y = dataFrame.extent.YMax
    df_proj_w = dataFrame.extent.width
    df_proj_h = dataFrame.extent.height
    # Ensure the coordinates are in the extent of the data frame.
    if pageX < df_page_x_min or pageX > df_page_x_max:
        pythonaddins.MessageBox('X coordinate is not within map portion of the page.', "Out of Bounds")
        return 0,0

    if pageY < df_page_y_min or pageY > df_page_y_max:
        pythonaddins.MessageBox('Y coordinate is not within map portion of the page.', "Out of Bounds")
        return 0,0

    # Scale the projected coordinates to map units from the lower left of the data frame.
    scale = dataFrame.scale/12 # Converting scale factor to feet per inch.
    map_x = df_min_x + ((pageX - df_page_x_min)*scale)
    map_y = df_min_y + ((pageY - df_page_y_min)*scale)

    return map_x,map_y
```

If you look further in this code, you will see each tool's class and the code to perform the selection. Each tool is taking the coordinates from the user's mouse click(s) and creating temporary geometry items. These geometry items are then used with the SelectLayerByLocation tool to select a set of features. One geometry item creates a polygon object from the user-drawn rectangle, and the other geometry item creates a point object from the single mouse click. This code may also be copied to your other add-in tools to allow for feature selections. The example in the following graphic shows the code for a point selection. The onMouseDownMap function captures the x,y coordinates and then checks to see whether the coordinates are in map units or page units. If necessary, the coordinates are sent to the convertCoords function and converted to map units. These are used to create a geometry object, which in turn is used for feature selections.

```
class Select_Point(object):
    """Implementation for SelectFeatures_addin.selpoint (Tool)"""
    def __init__(self):
        self.enabled = True
        self.shape = "NONE" # NONE means that the selection will be done with a single point.
        # The property self.shape can be set to "Line", "Circle", or "Rectangle" for interactive
        # shape drawing, and to activate the onLine/Polygon/Circle event sinks.
    def onMouseDownMap(self, x, y, button, shift):
        thisMap = arcpy.mapping.MapDocument("CURRENT")
        dataFrame = arcpy.mapping.ListDataFrames(thisMap)[0]
        page_x = x
        page_y = y

        # Check to see if the values are in page measurements.
        # If they are, then convert to projected coordinates.
        if page_x < 100:  # Assumes that the page is less than 100 inches wide.
        # The page coordinates are sent to the convertCoords function above,
        # and returned as map coordinates
            CN = convertCoords(page_x,page_y)
            print "The X coord is " + str(CN[0])
            print "The Y coord is " + str(CN[1])
        else:
            # If they aren't page units they are already projected,
            # and you can use them as captured.
            CN = page_x,page_y

        # Make point object for selections.
        pointGeom = arcpy.PointGeometry(arcpy.Point(CN[0],CN[1]), thisMap.activeDataFrame.spatialReference)
        # Get layer to select.
        selectLayer = arcpy.mapping.ListLayers(thisMap,"Parcels",dataFrame)[0]
        # Perform selection.
        arcpy.SelectLayerByLocation_management(selectLayer,"INTERSECT",pointGeom)
        arcpy.RefreshActiveView()
```

The code also contains an example of selecting with a rectangle. The process is basically the same, except that it involves more coordinates. This code can be copied with minor modifications to any of your projects in which you would like to make selections by point or rectangle.

When the wizard creates your add-in, it creates many files and folders to contain all the portions of the add-in. One of these files is the config.xml file, which contains code to manage the components of the add-in. Normally, this file is off-limits and should never be edited by the user. However, if you are careful and fully understand what the file contains, there are a few modifications that can be made. These modifications should be done carefully. One change is to alter the order in which the tools are displayed on a toolbar. By default, they appear in the order in which they were created. In the <Toolbars> section of the code, each tool is listed in its order of appearance, left to right. Moving a tool's reference line will alter where it appears on the toolbar. Shown in the graphic is the reference ID for the Select Rectangle tool. The entire line can be moved to a new location.

```
        <Toolbar caption="Select Features" category="Select
Features in Map" id="SelectFeatures_addin.selfeattoolbar"
showInitially="true"><Items>
<Tool refID="SelectFeatures_addin.selrect" />
<Tool refID="SelectFeatures_addin.selpoint" />
</Items></Toolbar>
        </Toolbars>
```

The other modification you can make is to add ArcGIS system tools to the toolbar. Look on the ArcGIS Resources website for Python for ArcGIS to find the list and IDs for commands in ArcMap, ArcCatalog, ArcScene, and ArcGlobe. These IDs can be used to add the buttons or tools to your custom toolbars or menus. As an example, to add buttons for switching from layout view to data view (or back), first look up the commands and get their program ID (ProgID), as shown:

This table contains the following information:
. Caption, name, and globally unique identifier (GUID) of the built-in toolbars in ArcMap.
. Caption, name, GUID, and parent of the built-in menus in ArcMap.
. Caption, name, command category (category in the Customize dialog box), GUID, SubType, parent, and description of the built-in commands that appear on the toolbars and menus in ArcMap.

Type	Caption	Name	Command Category	GUID (CLSID / ProgID)	Sub Type	Parent	Description
Menu	View	View_Menu	none	{119591C1-0255-11D2-8D20-080009EE4E51} esriArcMapUI.MxViewMenu	none	Main Menu	none
Command	Data View	View_Geographic	View	{65702489-A258-11D1-8740-0000F8751720} esriArcMapUI.GeographicViewCommand	none	View_Menu	Switches to data view
Command	Layout View	View_LayoutView	View	{6570248A-A258-11D1-8740-0000F8751720} esriArcMapUI.LayoutViewCommand	none	View_Menu	Switches to layout view

Then carefully add the line of code to the config.xml file to add the item. In the example shown in the following graphic, the two buttons handling the map view have been added along with the code to add the Identify tool. From the chart, also note that there are many geoprocessing tools that can be added this way. They will not, however, accept input from your code or provide data to other processing tasks in your code.

```xml
    <Toolbar caption="Select Features" category="Select
Features in Map" id="SelectFeatures_addin.selfeattoolbar"
showInitially="true"><Items>
<Tool refID="SelectFeatures_addin.selrect" />
<Tool refID="SelectFeatures_addin.selpoint" />
<Button refID="esriArcMapUI.GeographicViewCommand" />
<Button refID="esriArcMapUI.LayoutViewCommand" />
<Tool refID="esriControls.ControlsMapIdentifyTool" />
</Items></Toolbar>
    </Toolbars>
```

Build a complex Python add-in

1. Start the Python Add-In Wizard, and create a new project named Tutorial 5-3. **Add the toolbar and other items necessary to make the template for this project. Use the class names and IDs shown in the graphic, and provide tooltips and Help text for all the items.**

Item	Caption	Class Name	ID (Variable Name)	Task
Toolbar	Make Exhibit Maps		exbtoolbar	The main toolbar to hold the application items
Combo Box	Enter the Tax ID	Accept_TaxID	acceptid	The user can type the Tax ID number
Tool Palette	Selection Tools		seltools	Palette for the selection tools (Menu Style)
Tool	Select with Point	Select_Point	selpoint	The user can select a subject tract with a single mouse click
Tool	Select with Rectangle	Select_Rect	selrect	The user can select many subject tracts by drawing a rectangle
Combo Box	Enter the Map Title	Accept_Title	accepttitle	The user types the map title
Combo Box	Enter the Map Subtitle	Accept_Subtitle	acceptsubtitle	The user types the map subtitle
Menu	Print Menu		printmenu	Menu to hold the three print buttons
Button	Print Water Map	Print_Water	printwater	Sets the water layers and exports the map to a PDF
Button	Print Sewer Map	Print_Sewer	printsewer	Sets the sewer layers and exports the map to a PDF
Button	Print Storm Water Map	Print_StormWater	printstorm	Sets the storm water layers and exports the map to a PDF

With all the required items created, your template should resemble this graphic:

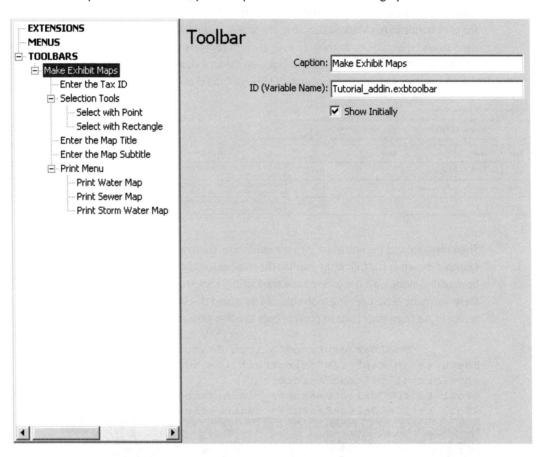

2.   **Save your project in the Python Add-In Wizard. Close the wizard.**

The framework for the toolbar is completed. If you have completed this process correctly, you will be able to add and edit the code associated with the menu without having to start over as in the example in tutorial 5-2. Next move on to adding code for the buttons. An instruction is given for each item on the toolbar to help you understand the code you will write. Pay attention to the enable property as you write the code for each item you have designed. You will want to disable the entry boxes for the title and subtitle until a valid parcel selection is made.

3.   **Navigate to the Tutorial 5-3 folder where you created the add-in. Edit the Tutorial_addin.py file created in the Install folder of your add-in.**

The first set of code to work with is the combo box for the tax ID entry. The _init_ function sets up the basic input parameters.

Because the combo box will accept only typed responses, the list of choices for the drop-down box will be empty, and the property editable will be set to True. The button will be enabled when it initializes, and you can set the width of the entry box to eight characters, as shown:

```
def __init__(self):
    self.items = []
    self.editable = True
    self.enabled = True
    self.dropdownWidth = 'W'
    self.width = 'WWWWWWWW'
```

Next write the code to accept the data entry and to find the parcel matching the entered tax ID number.

4. The data capture occurs in the onEnter function, so scroll to that part of the code for the Accept_TaxID class. The number entered is captured in the self object as self.value. Add a line of code to save this value in a variable named taxAccount, as shown:

```
def onEnter(self):
    # Store the entered value in a variable
    taxAccount = self.value
```

5. Create a layer object for the Parcels layer, from which you can do the selection. (Hint: set up the three objects for the map document, the data frame, and a specific layer in the table of contents, as shown.)

```
# Create a map document object for the currently open map document;
# a data frame list of the only data frame in this map document;
# a layer object of just the Parcels layer
thisMap = arcpy.mapping.MapDocument("CURRENT")
# Adding the index number at the end of the data frame object
# and layer object (with wildcard) creation
# means that you are working with only the first data frame
# and the layer called Parcels
myDF = arcpy.mapping.ListDataFrames(thisMap,"")[0]
myLayer = arcpy.mapping.ListLayers(thisMap,"Parcels",myDF)[0]
```

The last part of the onEnter code is to make the selection, pan to it, and enable the combo box to accept the map title.

6. Add the SelectLayerByAttribute tool to find where the field EKEY equals the tax ID entered by the user. If a valid selection is made, pan to that location, and enable the map title combo box, as shown. (Hint: check to make sure the selection is not equal to zero before trying to get the extent object, and pan the map.)

```
# Perform the selection
arcpy.SelectLayerByAttribute_management(myLayer,"NEW_SELECTION","EKEY = " + taxAccount)
# Check to see if an item is selected
# If so, pan to it and activate the Map Title combo box
if arcpy.GetCount_management(myLayer).getOutput(0) <> 0:
    # Get the extent of the selected feature
    lyrExtent = myLayer.getSelectedExtent()
    # Pan to the selected feature
    myDF.panToExtent(lyrExtent)
    # Use the global ID for the map title combo box to enable it
    accepttitle.enabled = True
```

7. When you have the code ready, create the add-in file, and install and test the file with the map document Tutorial 5-3 before moving on. A good test value for the tax ID is 40227464. If it is working correctly, the parcel should be selected, highlighted, and the map panned to that location.

The user has the option to interactively select the feature in the map rather than type a known tax ID. The classes Select_Point and Select_Rect hold the code for these tools.

8. Configure the two selection tools as shown in the SelectFeatures add-in described in this tutorial's "Scripting techniques" section. The Select_Point tool uses the onMouseDownMap handler, and the Select_Rect tool uses the onRectangle handler. Each tool should also test for valid selections, pan to the selected feature(s), and enable the combo box for the title entry.

9. Open the SelectFeatures_addin.py file in the folder C:\EsriPress\GISTPython\Python Add-Ins\ SelectFeatures\Install. Copy the convertCoords function (lines 4 through 38), and paste the function to the Tutorial addin.py file for your new toolbar. This function should go at the top of the script file, just after the import pythonaddins statement. Do not alter the convertCoords code.

10. From the SelectFeatures_addin.py file, copy the code from the onRectangle function in the Sel_Rectangle class (lines 45 through 85). Go to the Tutorial addin.py file for your new toolbar, and replace the onRectangle function in the Select_Rect class.

The template code you copied will handle the rectangle selection, but you must still set the _init_ function and add the code to the bottom of the selection to pan the map and enable the map title combo box.

11. In the _init_ function of the Select_Rect class, make sure that self.enabled is True, and set self.shape to Rectangle. Then add the code at the end of the onRectangle function to make sure at least one item is selected, pan the map to the extent of the selected features, and enable the map title entry. (Hint: Copy the last nine lines of the onEnter function in the Accept_TaxID class, and paste those lines at the end of the onRectangle function. Be careful to use the correct indentations, and check all variable names.)

Your turn

Configure the Select_Point class to allow the user to select a feature with a single point click. You can copy the code from the template SelectFeatures add-in, making sure to set the self.shape to NONE. Also, add the nine lines that will pan the map to the selected features and activate the map title entry. Notice how the coordinates for the clicked location are tested and, if necessary, sent to the convertCoords function and returned as map coordinates.

12. Take a moment to test the selection tools. Make sure the tool palette works as expected, the selections occur, the map pans, and the map title entry box is enabled. Remember that to test the add-in, you must save your code, exit ArcMap, run the makeaddin.py script, install the add-in file, restart ArcMap, and then open the map document Tutorial 5-3. When you are done testing the selection part of the script, close ArcMap.

Have the user type a title for the map. This box should initialize as disabled and will be enabled when selections are made using one of the other tools. This box uses the onEnter handler to accept what the user types in much the same way that the Accept_TaxID class does. When there is a valid entry, have the title in the map change and enable the subtitle entry box.

13. In your IDE, find the Accept_Title class, and set the _init_ parameters. These parameters should be the same as for the Accept_TaxID class, with the exception of the enabled property being set to False. This property being set to False will initialize the tool as disabled, but any of the selection tools will enable it when a proper selection is made.

14. Use the onEnter function to store the user's entry and change the map's title, as shown in the graphic. The process for this is similar to that in tutorial 4-1. After refreshing the map, use the combo box ID for the subtitle entry class to enable that combo box.

```
class Accept_Title(object):
    """Implementation for Tutorial_addin.accepttitle (ComboBox)"""
    def __init__(self):
        self.items = []
        self.editable = True
        self.enabled = False
        self.dropdownWidth = 'W'
        self.width = 'WWWWWWWW'
    def onSelChange(self, selection):
        pass
    def onEditChange(self, text):
        pass
    def onFocus(self, focused):
        pass
    def onEnter(self):
        mapTitle = self.value
        # Set up an object for the map document
        thisMap = arcpy.mapping.MapDocument("CURRENT")
        titleElement = arcpy.mapping.ListLayoutElements(thisMap, "TEXT_ELEMENT", "Map Title")[0]
        titleElement.text = mapTitle
        acceptsubtitle.enabled = True
        # Refresh the map
        arcpy.RefreshActiveView()
    def refresh(self):
        pass
```

Your turn

The last data entry is the subtitle. The ability to enter this item is disabled at initialization but is enabled when the map title is populated. Configure the Accept_Subtitle class to accept a valid entry, change the subtitle in the map elements, and refresh the map.

Once again you might want to test and troubleshoot your application. You should be able to use any one of the three methods of selecting features, and then set the title and subtitle.

You may have thought about disabling the print buttons until all the other actions are completed. Unfortunately, that is not possible with buttons or tools added to a menu. The items will not initialize until the menu is selected, so the objects on the menu cannot be controlled in the code.

But on a positive note, you can move on to writing the code for the print buttons. These buttons basically turn the correct layers on and off and then export the map to a PDF document. The code used in tutorial 3-2 can be copied and pasted to the add-in code, and with a little modification, the code can be made workable.

The first of the three print buttons is for the water map. This button turns on the group layer for the water utility data, the physical feature data, and the base group data. You may also want to check that the other group layers are turned off. Then export the map with the map title appended to the default name as **Water_Map_AI_**.

15. **In your MyExercises folder, open the RealtorExhibit.py file from tutorial 3-2. Copy the code that creates the map document object, the data frame object, and the list of layers from the table of contents. Then paste this code to the onClick function of the Print_Water class replacing the word *pass*, as shown:**

```
class Print_Water(object):
    """Implementation for Tutorial_addin.printwater (Button)"""
    def __init__(self):
        self.enabled = True
        self.checked = False
    def onClick(self):
        # Create a map document object for the currently open map document;
        # a data frame list of the only data frame in this map document;
        # a layer list object of all the layers
        thisMap = arcpy.mapping.MapDocument("CURRENT")
        myDF = arcpy.mapping.ListDataFrames(thisMap,"")[0]
        # Adding the index number at the end of the data frame object creation
        #  means that you are working with only the first data frame
        myLayers = arcpy.mapping.ListLayers(myDF)
```

16. In the RealtorExhibit.py file, find the line 'if "Water Utility Map" in maplist:. ' Starting with the next line after this, copy the code down to the line that creates the PDF document. Paste this code to the onClick function just below the other code. Make sure to use the ID of the map title item in the map's name, as shown:

```
arcpy.AddWarning("Creating the water map ...")
# Run through the layer groups and set visibility
for lyr in myLayers:
    if lyr.name == "Storm Water Utility Group":
        lyr.visible = "False"
    if lyr.name == "Water Utility Group":
        lyr.visible = "True"
    if lyr.name == "Sewer Utility Group":
        lyr.visible = "False"
    if lyr.name == "Physical Features Group":
        lyr.visible = "True"
    if lyr.name == "Base Group":
        lyr.visible = "True"
# Refresh the map
arcpy.RefreshActiveView()
arcpy.RefreshTOC()
# Output completed map to a PDF file
arcpy.mapping.ExportToPDF(thisMap,r"C:\EsriPress\GISTPython\MyExercises\Water_Map_AI_"\
    + accepttitle.value)
```

Your turn

Set up the print buttons for the sewer and storm water exhibit maps. The code for the Print_Sewer and Print_Storm Water classes can be copied from the RealtorExhibit.py file and modified for use.

17. Test the application using the map document Tutorial 5-3. Try creating the following map types for the sample locations that are shown in the chart with their tax ID numbers. For the interactive selections, pan around, and try random parcels located anywhere on the map.

Map Types	Title	Subtitle	Tax Account
All	Royal Oaks Estates	Block 5, Lot 8	2571323
Water, Sewer	Woodcreek Addition	Block D, Lot 14	3599809
Sewer, Storm Water	Somerset Place	Block C, Lot 19	2792826
Sewer, Storm Water, Water	Alexander Addition	Block 10, Lot 1	18007

This application uses all the more common components of Python add-ins, so it should serve as a great starting point for all the rest of the add-ins you write.

Exercise 5-3

Re-create the application you built in exercise 3-2 using Python add-ins. The scenario is that the city planner has seen the add-in for real estate agents and wants a similar application for property owner notification maps. He would like to be able to enter a tax account number or interactively select from the map and have the application automatically create a 200-foot buffer. Then he would like to zoom to that area and create a parcel map and a physical features map.

A sample map document Exercise 5-3 is provided. The key differences in this application are as follows:

- The user can enter more than one tax account number.
- The map is zoomed rather than panned.
- There are two data frames in this map document.

Name all the elements in the map document, and prepare it for automation. Then write the code and create a Python add-in application to accept an account number or interactive selection, and optionally a buffer distance (some cases may require more than a 200-foot buffer).

When the tool is finished running, you should have two maps for each case.

The first is a parcel map with these layers:
- Lot_Boundaries
- PlatIndex
- Blocks
- Lot Numbers
- ZoningDistricts
- Street Names
- The new buffer layer you create

The second is a physical features map with these layers:
- Lot_Boundaries
- Address Numbers
- Building Footprints
- Paved areas
- Creeks
- Bodies of Water
- Recreational Features
- Street Names
- The new buffer layer you create

(**Hint:** design the toolbar first, and then work on the code, reusing as much code as possible from exercise 3-2.)

Here are some sample cases for testing, or use the interactive selections, and select any parcels:

Tax Account Numbers	Map Title	Case Type	Property Description 1	Property Description 2
1784102	RZ-04-2014	Rezone	SW Mills Addition	Block 2, Lot 4R
41433068	SP-12-2014	Site Plan	Fountain Center Addition	Lot 2
653764	PL-09-2014	Plat	Cresthaven Addition	Block 1, Lot A1
2024888	VR-11-2014	Variance	Oakland Estates	Block 2, Lot 2

Tutorial 5-3 review

The ability to interact with the map adds a great deal of convenience to your applications and enhances the user experience. Although not done here, you could add your own tool palette of zooming and panning tools to the add-in toolbar. You can also add a drop-down menu with other system tools that might be useful within the application.

The Python add-Ins framework has some limitations, such as ensuring that the toolbar design is right the first time and restarting ArcMap each time you test the code. However, users will not experience these limitations, so the result will justify the work you put into your coding and debugging efforts.

Study questions

1. What other user interactions with the map document could you monitor and use as a means of selection?

2. What other monitored events could you use to collect user data from the mouse or keyboard?

3. What effect would different map units or different coordinate systems have on the interactive selection routine shown in this tutorial?

There are many online resources for learning more about Python add-Ins on the ArcGIS Resources website, resources.arcgis.com, and the Esri Training website, training.esri.com. The Python community on the ArcGIS Resources website has many articles and blogs concerning Python programming, links to other training opportunities, and a discussion forum where you can post questions about your code and get help finding solutions. The Esri Training website offers a variety of Python training formats from conventional classroom instruction to free online webcasts.

APPENDIX A

Using an IDE for Python scripting

Many integrated development environment (IDE) applications are available for programming in Python. Some IDEs are free to download and use, and other IDEs are commercially available. The decision on which one to use ultimately rests with you and will depend on your needs and budget. There are three free, widely available options discussed here, mostly to show how these IDEs can be used in Python scripting for ArcGIS.

Choosing an IDE

Whichever IDE you choose should be made the default in ArcGIS. With your chosen IDE set as the default, any time you begin editing a Python script tool or a Python toolbox in ArcMap or ArcCatalog, your default IDE will open. To set the default IDE program, go to the Geoprocessing menu in either ArcMap or ArcCatalog and click Geoprocessing Options. In the Script Tool Editor/Debugger box, browse to and select the IDE executable file you wish to use, as shown:

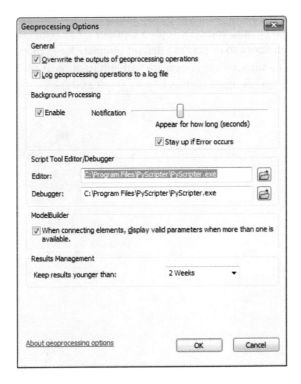

PythonWin

PythonWin is a good, Windows-based IDE, which works well with ArcGIS. It tracks variables and offers automatic code completion prompts when you are writing code. It also tracks and maintains the indentations in your code. The drawbacks of PythonWin include its restriction to run only in the Windows operating system and its limited debugging capabilities. It can be downloaded and used for free.

IDLE

IDLE (integrated development environment) comes free with the Python installation files included with ArcGIS. Its strength is that it runs on a variety of platforms, but it is limited in functionality. It does not include debugging tools or indentation handling, which is a shortcoming when coding in ArcGIS. It is a very basic IDE and good for beginners, but most users will move on to a more feature-rich IDE as they gain code-writing experience.

PyScripter

PyScripter is a solid IDE that can be downloaded for free. It has robust debugging tools, good automatic code completion functionality, and tracks indentations. In addition to its basic features, it also has many features that assist in coding. For example, when you use a bracket, parenthesis, quotation mark, or many of the other characters used to start and stop an entry, PyScripter automatically adds and tracks the closing character. This feature is useful when nesting these characters and helps to make sure that all entries are properly closed.

PyScripter also allows the user to set up a file template, which for ArcGIS scripts can include your name and date as well as import arcpy and other common setup commands. To edit the default script template, start PyScripter, go to the main menu, and click Tools > Options > File Templates. In the File Templates dialog box, select Python Script to display the default template for Python scripts. The template contains a header with code to automatically populate such information as the author and date, below which you can add ArcGIS-specific code such as import arcpy, as shown:

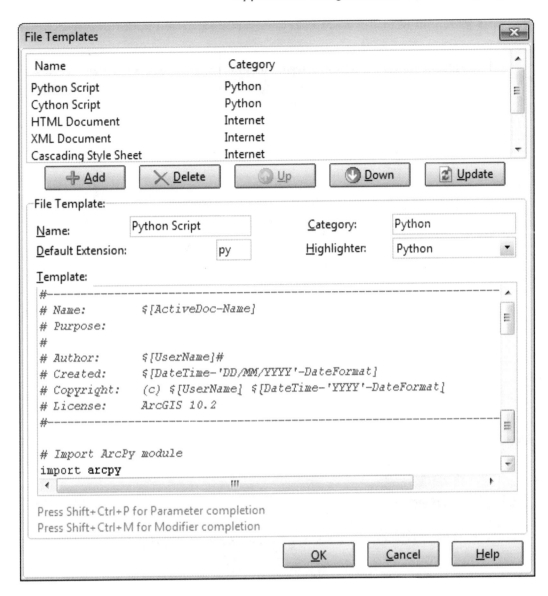

IDE for Python toolboxes

Python toolboxes in ArcGIS use a file with a .pyt extension. In order for the IDE to apply scripting features to the code, such as syntax highlighting, code completion prompts, and tracking indent levels, the IDE must be configured to recognize this extension.

In PythonWin

PythonWin automatically sees a .pyt file as a script and applies all coding features to the file.

In IDLE

The following steps enable syntax highlighting on a Python toolbox that is being edited in IDLE:

1. Start Windows Explorer, and navigate to the idlelib folder for the install of IDLE. By default, this location should be similar to C:\Python27\ArcGIS10.2\Lib\idlelib.

2. Open the EditorWindow.py file for editing in a Python IDE or Notepad.

3. Locate the following lines of code within the EditorWindow.py file (search for os.path.normcase), as shown:

```
def ispythonsource(self, filename):
    if not filename or os.path.isdir(filename):
        return True
    base, ext = os.path.splitext(os.path.basename(filename))
    if os.path.normcase(ext) in (".py", ".pyw"):
        return True
```

4. Add **.pyt** to the list of extensions. The updated EditorWindow.py file should look similar to this:

```
def ispythonsource(self, filename):
    if not filename or os.path.isdir(filename):
        return True
    base, ext = os.path.splitext(os.path.basename(filename))
    if os.path.normcase(ext) in (".py", ".pyw", ".pyt"):
        return True
```

5. Save the changes to the EditorWindow.py file.

In PyScripter

The following steps enable the coding features on a Python toolbox (.pyt) that is being edited in PyScripter:

1. Start PyScripter.

2. Click Tools > Options > IDE Options.

3. Scroll down to the File Filters section of the IDE Options dialog box.

4. Add a semicolon and ***.pyt** to the Python files for Open dialog Python filter. Note that this is added in two places on the entry line, as shown:

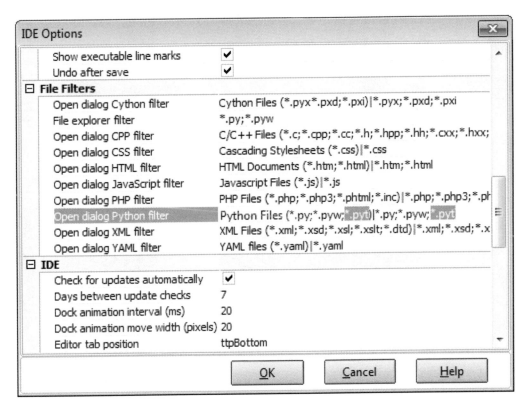

5. Click OK.

APPENDIX B

Tool index

Software tool/concept, tutorial(s) in which it appears

APPENDIX C

Data source credits

\\EsriPress\GISTPython\Data\City of Oleander.gdb\FireBoxMaps\FireBoxMap_100, courtesy of the City of Euless.

\\EsriPress\GISTPython\Data\City of Oleander.gdb\FireBoxMaps\FireBoxMap_101, courtesy of the City of Euless.

\\EsriPress\GISTPython\Data\City of Oleander.gdb\FireBoxMaps\FireBoxMap_102, courtesy of the City of Euless.

\\EsriPress\GISTPython\Data\City of Oleander.gdb\FireBoxMaps\FireBoxMap_103, courtesy of the City of Euless.

\\EsriPress\GISTPython\Data\City of Oleander.gdb\FireBoxMaps\FireBoxMap_104, courtesy of the City of Euless.

\\EsriPress\GISTPython\Data\City of Oleander.gdb\FireBoxMaps\FireBoxMap_105, courtesy of the City of Euless.

\\EsriPress\GISTPython\Data\City of Oleander.gdb\FireBoxMaps\FireBoxMap_106, courtesy of the City of Euless.

\\EsriPress\GISTPython\Data\City of Oleander.gdb\FireBoxMaps\FireBoxMap_107, courtesy of the City of Euless.

\\EsriPress\GISTPython\Data\City of Oleander.gdb\FireBoxMaps\FireBoxMap_108, courtesy of the City of Euless.

\\EsriPress\GISTPython\Data\City of Oleander.gdb\FireBoxMaps\FireBoxMap_109, courtesy of the City of Euless.

\\EsriPress\GISTPython\Data\City of Oleander.gdb\FireBoxMaps\FireBoxMap_110, courtesy of the City of Euless.

\\EsriPress\GISTPython\Data\City of Oleander.gdb\FireBoxMaps\FireBoxMap_111, courtesy of the City of Euless.

\\EsriPress\GISTPython\Data\City of Oleander.gdb\FireBoxMaps\FireBoxMap_112, courtesy of the City of Euless.

\\EsriPress\GISTPython\Data\City of Oleander.gdb\FireBoxMaps\FireBoxMap_113, courtesy of the City of Euless.

\\EsriPress\GISTPython\Data\City of Oleander.gdb\FireBoxMaps\FireBoxMap_114, courtesy of the City of Euless.

\\EsriPress\GISTPython\Data\City of Oleander.gdb\FireBoxMaps\FireBoxMap_115, courtesy of the City of Euless.

\\EsriPress\GISTPython\Data\City of Oleander.gdb\FireBoxMaps\FireBoxMap_116, courtesy of the City of Euless.

\\EsriPress\GISTPython\Data\City of Oleander.gdb\FireBoxMaps\FireBoxMap_117, courtesy of the City of Euless.

\\EsriPress\GISTPython\Data\City of Oleander.gdb\FireBoxMaps\FireBoxMap_200, courtesy of the City of Euless.

\\EsriPress\GISTPython\Data\City of Oleander.gdb\FireBoxMaps\FireBoxMap_201, courtesy of the City of Euless.

\\EsriPress\GISTPython\Data\City of Oleander.gdb\FireBoxMaps\FireBoxMap_202, courtesy of the City of Euless.

\\EsriPress\GISTPython\Data\City of Oleander.gdb\FireBoxMaps\FireBoxMap_203, courtesy of the City of Euless.

\\EsriPress\GISTPython\Data\City of Oleander.gdb\FireBoxMaps\FireBoxMap_204, courtesy of the City of Euless.

\\EsriPress\GISTPython\Data\City of Oleander.gdb\FireBoxMaps\FireBoxMap_205, courtesy of the City of Euless.

\\EsriPress\GISTPython\Data\City of Oleander.gdb\FireBoxMaps\FireBoxMap_206, courtesy of the City of Euless.

\\EsriPress\GISTPython\Data\City of Oleander.gdb\FireBoxMaps\FireBoxMap_207e, courtesy of the City of Euless.

\\EsriPress\GISTPython\Data\City of Oleander.gdb\FireBoxMaps\FireBoxMap_207w, courtesy of the City of Euless.

\\EsriPress\GISTPython\Data\City of Oleander.gdb\FireBoxMaps\FireBoxMap_208, courtesy of the City of Euless.

\\EsriPress\GISTPython\Data\City of Oleander.gdb\FireBoxMaps\FireBoxMap_209, courtesy of the City of Euless.

\\EsriPress\GISTPython\Data\City of Oleander.gdb\FireBoxMaps\FireBoxMap_210, courtesy of the City of Euless.

\\EsriPress\GISTPython\Data\City of Oleander.gdb\FireBoxMaps\FireBoxMap_300, courtesy of the City of Euless.

\\EsriPress\GISTPython\Data\City of Oleander.gdb\FireBoxMaps\FireBoxMap_301, courtesy of the City of Euless.

\\EsriPress\GISTPython\Data\City of Oleander.gdb\FireBoxMaps\FireBoxMap_302, courtesy of the City of Euless.

\\EsriPress\GISTPython\Data\City of Oleander.gdb\FireBoxMaps\FireBoxMap_303, courtesy of the City of Euless.

\\EsriPress\GISTPython\Data\City of Oleander.gdb\FireBoxMaps\FireBoxMap_304, courtesy of the City of Euless.

\\EsriPress\GISTPython\Data\City of Oleander.gdb\FireBoxMaps\FireBoxMap_305, courtesy of the City of Euless.

\\EsriPress\GISTPython\Data\City of Oleander.gdb\FireBoxMaps\FireBoxMap_306, courtesy of the City of Euless.

\\EsriPress\GISTPython\Data\City of Oleander.gdb\FireBoxMaps\FireBoxMap_307, courtesy of the City of Euless.

\\EsriPress\GISTPython\Data\City of Oleander.gdb\FireBoxMaps\FireBoxMap_308, courtesy of the City of Euless.

\\EsriPress\GISTPython\Data\City of Oleander.gdb\FireBoxMaps\FireBoxMap_309, courtesy of the City of Euless.

\\EsriPress\GISTPython\Data\City of Oleander.gdb\FireBoxMaps\FireBoxMap_318, courtesy of the City of Euless.

\\EsriPress\GISTPython\Data\City of Oleander.gdb\FireBoxMaps\FireBoxMap_319, courtesy of the City of Euless.

\\EsriPress\GISTPython\Data\City of Oleander.gdb\FireBoxMaps\FireBoxMap_320, courtesy of the City of Euless.

\\EsriPress\GISTPython\Data\City of Oleander.gdb\FireBoxMaps\FireBoxMap_321, courtesy of the City of Euless.

\\EsriPress\GISTPython\Data\City of Oleander.gdb\Planimetrics\AnalysisCreeks, courtesy of the City of Euless.

\\EsriPress\GISTPython\Data\City of Oleander.gdb\Planimetrics\BldgFootprints, courtesy of the City of Euless.

\\EsriPress\GISTPython\Data\City of Oleander.gdb\Planimetrics\BodiesOfWater, courtesy of the City of Euless.

\\EsriPress\GISTPython\Data\City of Oleander.gdb\Planimetrics\Contours, courtesy of the City of Euless.

\\EsriPress\GISTPython\Data\City of Oleander.gdb\Planimetrics\Creeks, courtesy of the City of Euless.

\\EsriPress\GISTPython\Data\City of Oleander.gdb\Planimetrics\ElecUtil, courtesy of the City of Euless.

\\EsriPress\GISTPython\Data\City of Oleander.gdb\Planimetrics\Fences, courtesy of the City of Euless.

\\EsriPress\GISTPython\Data\City of Oleander.gdb\Planimetrics\FWAparts, courtesy of the City of Euless.

\\EsriPress\GISTPython\Data\City of Oleander.gdb\Planimetrics\PavedAreas, courtesy of the City of Euless.

\\EsriPress\GISTPython\Data\City of Oleander.gdb\Planimetrics\PavingPolygons, courtesy of the City of Euless.

\\EsriPress\GISTPython\Data\City of Oleander.gdb\Planimetrics\RecFea, courtesy of the City of Euless.

\\EsriPress\GISTPython\Data\City of Oleander.gdb\Planimetrics\TreeMass, courtesy of the City of Euless.

\\EsriPress\GISTPython\Data\City of Oleander.gdb\Well_Data\BC_ASSOC_1H_Path, created by the author.

\\EsriPress\GISTPython\Data\City of Oleander.gdb\Well_Data\BC_ASSOC_2H_Path, created by the author.

\\EsriPress\GISTPython\Data\City of Oleander.gdb\Well_Data\BC_ASSOC_3H_Path, created by the author.

\\EsriPress\GISTPython\Data\City of Oleander.gdb\Well_Data\BC_ASSOC_4H_Path, created by the author.

\\EsriPress\GISTPython\Data\City of Oleander.gdb\Well_Data\BC_ASSOC_5H_Path, created by the author.

\\EsriPress\GISTPython\Data\City of Oleander.gdb\Well_Data\BC_ASSOC_6H_Path, created by the author.

\\EsriPress\GISTPython\Data\City of Oleander.gdb\Well_Data\BC_South_1H_Path, created by the author.

\\EsriPress\GISTPython\Data\City of Oleander.gdb\Well_Data\BC_South_2H_Path, created by the author.

\\EsriPress\GISTPython\Data\City of Oleander.gdb\Well_Data\BC_South_3H_Path, created by the author.

\\EsriPress\GISTPython\Data\City of Oleander.gdb\Well_Data\WellSites, created by the author.

\\EsriPress\GISTPython\Data\City of Oleander.gdb\Address_NumbersAnno, courtesy of the City of Euless.

\\EsriPress\GISTPython\Data\City of Oleander.gdb\Address_NumbersAnno_1, courtesy of the City of Euless.

\\EsriPress\GISTPython\Data\City of Oleander.gdb\Blocks, courtesy of the City of Euless.

\\EsriPress\GISTPython\Data\City of Oleander.gdb\BookmobileLocations, created by the author.

\\EsriPress\GISTPython\Data\City of Oleander.gdb\BuildingFootprints, courtesy of the City of Euless.

\\EsriPress\GISTPython\Data\City of Oleander.gdb\Calls_for_service_2010, courtesy of the City of Euless.

\\EsriPress\GISTPython\Data\City of Oleander.gdb\Calls_for_service_2012, courtesy of the City of Euless.

\\EsriPress\GISTPython\Data\City of Oleander.gdb\City_Area, courtesy of the City of Euless.

\\EsriPress\GISTPython\Data\City of Oleander.gdb\CityLimitLine, courtesy of the City of Euless.

\\EsriPress\GISTPython\Data\City of Oleander.gdb\ComPlan, courtesy of the City of Euless.

\\EsriPress\GISTPython\Data\City of Oleander.gdb\FireBoxMap, courtesy of the City of Euless.

\\EsriPress\GISTPython\Data\City of Oleander.gdb\FireBoxMapAnno, courtesy of the City of Euless.

\\EsriPress\GISTPython\Data\City of Oleander.gdb\GeneralZoningDistricts, courtesy of the City of Euless.

\\EsriPress\GISTPython\Data\City of Oleander.gdb\Lot_Boundaries, courtesy of the City of Euless.

\\EsriPress\GISTPython\Data\City of Oleander.gdb\Monument, courtesy of the City of Euless.

\\EsriPress\GISTPython\Data\City of Oleander.gdb\Ownership, courtesy of the City of Euless.

\\EsriPress\GISTPython\Data\City of Oleander.gdb\Parcels, courtesy of the City of Euless.

\\EsriPress\GISTPython\Data\City of Oleander.gdb\PlatIndex, courtesy of the City of Euless.

\\EsriPress\GISTPython\Data\City of Oleander.gdb\PrivateRoads, courtesy of the City of Euless.

\\EsriPress\GISTPython\Data\City of Oleander.gdb\RecycleZones, courtesy of the City of Euless.

\\EsriPress\GISTPython\Data\City of Oleander.gdb\Street_Centerlines, courtesy of the City of Euless.

\\EsriPress\GISTPython\Data\City of Oleander.gdb\StreetLights, courtesy of the City of Euless.

\\EsriPress\GISTPython\Data\City of Oleander.gdb\VacantIndex, courtesy of the City of Euless.

\\EsriPress\GISTPython\Data\City of Oleander.gdb\ZoningBoundaries, courtesy of the City of Euless.

\\EsriPress\GISTPython\Data\City of Oleander.gdb\ZoningDistricts, courtesy of the City of Euless.

\\EsriPress\GISTPython\Data\FireDepartment.gdb\BldgFootprints, courtesy of the City of Euless.

\\EsriPress\GISTPython\Data\FireDepartment.gdb\FireBoxMap_0, courtesy of the City of Euless.

\\EsriPress\GISTPython\Data\FireDepartment.gdb\FireBoxMap_1, courtesy of the City of Euless.

\\EsriPress\GISTPython\Data\FireDepartment.gdb\FireBoxMap_10, courtesy of the City of Euless.

\\EsriPress\GISTPython\Data\FireDepartment.gdb\FireBoxMap_11, courtesy of the City of Euless.

\\EsriPress\GISTPython\Data\FireDepartment.gdb\FireBoxMap_12, courtesy of the City of Euless.

\\EsriPress\GISTPython\Data\FireDepartment.gdb\FireBoxMap_13, courtesy of the City of Euless.

\\EsriPress\GISTPython\Data\FireDepartment.gdb\FireBoxMap_14, courtesy of the City of Euless.

\\EsriPress\GISTPython\Data\FireDepartment.gdb\FireBoxMap_15, courtesy of the City of Euless.

\\EsriPress\GISTPython\Data\FireDepartment.gdb\FireBoxMap_16, courtesy of the City of Euless.

\\EsriPress\GISTPython\Data\FireDepartment.gdb\FireBoxMap_17, courtesy of the City of Euless.

\\EsriPress\GISTPython\Data\FireDepartment.gdb\FireBoxMap_18, courtesy of the City of Euless.

\\EsriPress\GISTPython\Data\FireDepartment.gdb\FireBoxMap_19, courtesy of the City of Euless.

\\EsriPress\GISTPython\Data\FireDepartment.gdb\FireBoxMap_2, courtesy of the City of Euless.

\\EsriPress\GISTPython\Data\FireDepartment.gdb\FireBoxMap_20, courtesy of the City of Euless.

\\EsriPress\GISTPython\Data\FireDepartment.gdb\FireBoxMap_21, courtesy of the City of Euless.

\\EsriPress\GISTPython\Data\FireDepartment.gdb\FireBoxMap_22, courtesy of the City of Euless.

\\EsriPress\GISTPython\Data\FireDepartment.gdb\FireBoxMap_23, courtesy of the City of Euless.

\\EsriPress\GISTPython\Data\FireDepartment.gdb\FireBoxMap_24, courtesy of the City of Euless.

\\EsriPress\GISTPython\Data\FireDepartment.gdb\FireBoxMap_25, courtesy of the City of Euless.

\\EsriPress\GISTPython\Data\FireDepartment.gdb\FireBoxMap_26, courtesy of the City of Euless.

\\EsriPress\GISTPython\Data\FireDepartment.gdb\FireBoxMap_27, courtesy of the City of Euless.

\\EsriPress\GISTPython\Data\FireDepartment.gdb\FireBoxMap_28, courtesy of the City of Euless.

\\EsriPress\GISTPython\Data\FireDepartment.gdb\FireBoxMap_29, courtesy of the City of Euless.

\\EsriPress\GISTPython\Data\FireDepartment.gdb\FireBoxMap_3, courtesy of the City of Euless.

\\EsriPress\GISTPython\Data\FireDepartment.gdb\FireBoxMap_30, courtesy of the City of Euless.

\\EsriPress\GISTPython\Data\FireDepartment.gdb\FireBoxMap_31, courtesy of the City of Euless.

\\EsriPress\GISTPython\Data\FireDepartment.gdb\FireBoxMap_32, courtesy of the City of Euless.

\\EsriPress\GISTPython\Data\FireDepartment.gdb\FireBoxMap_33, courtesy of the City of Euless.

\\EsriPress\GISTPython\Data\FireDepartment.gdb\FireBoxMap_34, courtesy of the City of Euless.

\\EsriPress\GISTPython\Data\FireDepartment.gdb\FireBoxMap_35, courtesy of the City of Euless.

\\EsriPress\GISTPython\Data\FireDepartment.gdb\FireBoxMap_36, courtesy of the City of Euless.

\\EsriPress\GISTPython\Data\FireDepartment.gdb\FireBoxMap_37, courtesy of the City of Euless.

\\EsriPress\GISTPython\Data\FireDepartment.gdb\FireBoxMap_38, courtesy of the City of Euless.

\\EsriPress\GISTPython\Data\FireDepartment.gdb\FireBoxMap_39, courtesy of the City of Euless.

\\EsriPress\GISTPython\Data\FireDepartment.gdb\FireBoxMap_4, courtesy of the City of Euless.

\\EsriPress\GISTPython\Data\FireDepartment.gdb\FireBoxMap_40, courtesy of the City of Euless.

\\EsriPress\GISTPython\Data\FireDepartment.gdb\FireBoxMap_41, courtesy of the City of Euless.

\\EsriPress\GISTPython\Data\FireDepartment.gdb\FireBoxMap_42, courtesy of the City of Euless.

\\EsriPress\GISTPython\Data\FireDepartment.gdb\FireBoxMap_43, courtesy of the City of Euless.

\\EsriPress\GISTPython\Data\FireDepartment.gdb\FireBoxMap_5, courtesy of the City of Euless.

\\EsriPress\GISTPython\Data\FireDepartment.gdb\FireBoxMap_6, courtesy of the City of Euless.

\\EsriPress\GISTPython\Data\FireDepartment.gdb\FireBoxMap_7, courtesy of the City of Euless.

\\EsriPress\GISTPython\Data\FireDepartment.gdb\FireBoxMap_8, courtesy of the City of Euless.

\\EsriPress\GISTPython\Data\FireDepartment.gdb\FireBoxMap_9, courtesy of the City of Euless.

\\EsriPress\GISTPython\Data\FireDepartment.gdb\OFD_Run_Data_Apr_2010, courtesy of the City of Euless.

\\EsriPress\GISTPython\Data\FireDepartment.gdb\OFD_Run_Data_Aug_2010, courtesy of the City of Euless.

\\EsriPress\GISTPython\Data\FireDepartment.gdb\OFD_Run_Data_Dec_2010, courtesy of the City of Euless.

\\EsriPress\GISTPython\Data\FireDepartment.gdb\OFD_Run_Data_Feb_2010, courtesy of the City of Euless.

\\EsriPress\GISTPython\Data\FireDepartment.gdb\OFD_Run_Data_Jan_2010, courtesy of the City of Euless.

\\EsriPress\GISTPython\Data\FireDepartment.gdb\OFD_Run_Data_Jul_2010, courtesy of the City of Euless.

\\EsriPress\GISTPython\Data\FireDepartment.gdb\OFD_Run_Data_Jun_2010, courtesy of the City of Euless.

\\EsriPress\GISTPython\Data\FireDepartment.gdb\OFD_Run_Data_Mar_2010, courtesy of the City of Euless.

\\EsriPress\GISTPython\Data\FireDepartment.gdb\OFD_Run_Data_May_2010, courtesy of the City of Euless.

\\EsriPress\GISTPython\Data\FireDepartment.gdb\OFD_Run_Data_Nov_2010, courtesy of the City of Euless.

\\EsriPress\GISTPython\Data\FireDepartment.gdb\OFD_Run_Data_Oct_2010, courtesy of the City of Euless.

\\EsriPress\GISTPython\Data\FireDepartment.gdb\OFD_Run_Data_Sep_2010, courtesy of the City of Euless.

\\EsriPress\GISTPython\Data\FireDepartment.gdb\OFD_Run_Data_TST_2010, courtesy of the City of Euless.

\\EsriPress\GISTPython\Data\OleanderOwnership.gdb\Elm_Fork_Addition, created by the author.

\\EsriPress\GISTPython\Data\OleanderOwnership.gdb\FireRuns2010, courtesy of the City of Euless.

\\EsriPress\GISTPython\Data\Sample Data.gdb\arbordaze2009tents, courtesy of the City of Euless.

\\EsriPress\GISTPython\Data\Sample Data.gdb\complan, courtesy of the City of Euless.

\\EsriPress\GISTPython\Data\Sample Data.gdb\libsprk, courtesy of the City of Euless.

\\EsriPress\GISTPython\Data\Sample Data.gdb\ROW_And_Easements, courtesy of the City of Euless.

\\EsriPress\GISTPython\Data\Sample Data.gdb\sprinklerunit, courtesy of the City of Euless.

\\EsriPress\GISTPython\Data\Sample Data.gdb\ZIPCODES_poly, courtesy of the City of Euless.

\\EsriPress\GISTPython\Data\SewerMaps.gdb\SewerMaps\SewerLin, courtesy of the City of Euless.

\\EsriPress\GISTPython\Data\SewerMaps.gdb\SewerMaps\SewerNod, courtesy of the City of Euless.

\\EsriPress\GISTPython\Data\SewerMaps.gdb\SewerMaps\TRAMeteringStations, courtesy of the City of Euless.

\\EsriPress\GISTPython\Data\SewerMaps.gdb\UtilityGrid300Scale, courtesy of the City of Euless.

\\EsriPress\GISTPython\Data\StormDrainUtility.gdb\Storm_Drains\Fixtures, courtesy of the City of Euless.

\\EsriPress\GISTPython\Data\StormDrainUtility.gdb\Storm_Drains\MainLat, courtesy of the City of Euless.

\\EsriPress\GISTPython\Data\WaterUtility.gdb\Cycles, courtesy of the City of Euless.

\\EsriPress\GISTPython\Data\WaterUtility.gdb\DistLateral, courtesy of the City of Euless.

\\EsriPress\GISTPython\Data\WaterUtility.gdb\DistMains, courtesy of the City of Euless.

\\EsriPress\GISTPython\Data\WaterUtility.gdb\EffluentWaterContractArea, courtesy of the City of Euless.

\\EsriPress\GISTPython\Data\WaterUtility.gdb\Fittings, courtesy of the City of Euless.

\\EsriPress\GISTPython\Data\WaterUtility.gdb\HydLaterals, courtesy of the City of Euless.

\\EsriPress\GISTPython\Data\WaterUtility.gdb\SamplingStations, courtesy of the City of Euless.

\\EsriPress\GISTPython\Data\2010 Run Data.xlsx\'2010 Run Data$,' courtesy of the City of Euless.

\\EsriPress\GISTPython\Data\2010 Run Data.xlsx.xml, courtesy of the City of Euless.

\\EsriPress\GISTPython\Data\2011 Run Data.xlsx\'2011 Run Data$,' courtesy of the City of Euless.

\\EsriPress\GISTPython\Data\ArcMap_MXT_File16.png, provided by Esri with the Python Add-In Wizard.

\\EsriPress\GISTPython\Data\CountRoutine.py, created by the author.

\\EsriPress\GISTPython\Data\DescribeObject.py, created by the author.

\\EsriPress\GISTPython\Data\ElementCircle16.png, provided by Esri with the Python Add-In Wizard.

\\EsriPress\GISTPython\Data\ElementLine16.png, provided by Esri with the Python Add-In Wizard.

\\EsriPress\GISTPython\Data\ElementRectangle16.png, provided by Esri with the Python Add-In Wizard.

\\EsriPress\GISTPython\Data\LocationGetPoint16.png, provided by Esri with the Python Add-In Wizard.

\\EsriPress\GISTPython\Data\Map1.png, provided by Esri with the Python Add-In Wizard.

\\EsriPress\GISTPython\Data\Map16.png, provided by Esri with the Python Add-In Wizard.

\\EsriPress\GISTPython\Data\Map2.png, provided by Esri with the Python Add-In Wizard.

\\EsriPress\GISTPython\Data\Map3.png, provided by Esri with the Python Add-In Wizard.

\\EsriPress\GISTPython\Data\MapSheet.png, provided by Esri with the Python Add-In Wizard.

\\EsriPress\GISTPython\Data\MapWithWrench16.png, provided by Esri with the Python Add-In Wizard.

\\EsriPress\GISTPython\Data\NauticalClearScaleBand.png, provided by Esri with the Python Add-In Wizard.

\\EsriPress\GISTPython\Data\NauticalSetScaleBand.png, provided by Esri with the Python Add-In Wizard.

\\EsriPress\GISTPython\Data\OleanderTX.jpg, created by the author.

\\EsriPress\GISTPython\Data\OleanderTX.png, created by the author.

\\EsriPress\GISTPython\Data\print.png, provided by Esri with the Python Add-In Wizard.

\\EsriPress\GISTPython\Data\Tutorial 4-1.txt, created by the author.

\\EsriPress\GISTPython\Data\Tutorial 4-2.txt, created by the author.

\\EsriPress\GISTPython\Data\Tutorial 4-3.txt, created by the author.

\\EsriPress\GISTPython\Data\Tutorial 5-1.txt, created by the author.

\\EsriPress\GISTPython\Data\Tutorial 5-2.txt, created by the author.

\\EsriPress\GISTPython\Maps, all MXD files created by the author for use with tutorials and exercises.

\\EsriPress\GISTPython\MyExercises\Scratch\Temporary Storage.gdb\BufferTemplate, created by the author.

\\EsriPress\GISTPython\MyExercises\Scratch\Temporary Storage.gdb\SelectionBuffer, created by the author.

\\EsriPress\GISTPython\MyExercises\Scratch\Temporary Storage.gdb\SiteTemp, created by the author.

\\EsriPress\GISTPython\MyExercises\MyAnswers.gdb, created by the author.

\\EsriPress\GISTPython\MyExercises\BufferTemp.lyr, created by the author.

\\EsriPress\GISTPython\MyExercises\Property Value 2009.lyr, created by the author.

\\EsriPress\GISTPython\MyExercises\Property Value 2010.lyr, created by the author.

\\EsriPress\GISTPython\MyExercises\Property Value 2011.lyr, created by the author.

\\EsriPress\GISTPython\MyExercises\Property Value 2012.lyr, created by the author.

\\EsriPress\GISTPython\Python Add-Ins\SelectFeatures\Images\ElementCircle16.png, provided by Esri with the Python Add-In Wizard.

\\EsriPress\GISTPython\Python Add-Ins\SelectFeatures\Images\ElementLine16.png, provided by Esri with the Python Add-In Wizard.

\\EsriPress\GISTPython\Python Add-Ins\SelectFeatures\Images\ElementRectangle16.png, provided by Esri with the Python Add-In Wizard.

\\EsriPress\GISTPython\Python Add-Ins\SelectFeatures\Images\LocationGetPoint16.png, provided by Esri with the Python Add-In Wizard.

\\EsriPress\GISTPython\Python Add-Ins\SelectFeatures\Install\SelectFeatures.py, created by the author.

\\EsriPress\GISTPython\Python Add-Ins\SelectFeatures\config.xml, provided by Esri with the Python Add-in Wizard.

\\EsriPress\GISTPython\Python Add-Ins\SelectFeatures\config2.xml, provided by Esri with the Python Add-in Wizard.

\\EsriPress\GISTPython\Python Add-Ins\SelectFeatures\makeaddin.py, provided by Esri with the Python Add-in Wizard.

\\EsriPress\GISTPython\Python Add-Ins\SelectFeatures\README.txt, provided by Esri with the Python Add-in Wizard.

\\EsriPress\GISTPython\Python Add-Ins\SelectFeatures\ResourceLinks.txt, provided by Esri with the Python Add-in Wizard.

\\EsriPress\GISTPython\Python Add-Ins\SelectFeatures\SelectFeatures.esriaddin, created by the author.

\\EsriPress\GISTPython\Python Add-Ins\SelectFeatures\SelectFeaturesView.esriaddin, created by the author.

\\EsriPress\GISTPython\Python Add-Ins\SelectNCount\Images\Map1.png, provided by Esri with the Python Add-In Wizard.

\\EsriPress\GISTPython\Python Add-Ins\SelectNCount\Install\SelectNCount_addin.py, created by the author.

\\EsriPress\GISTPython\Python Add-Ins\SelectNCount\config.xml, provided by Esri with the Python Add-in Wizard.

\\EsriPress\GISTPython\Python Add-Ins\SelectNCount\makeaddin.py, provided by Esri with the Python Add-in Wizard.

\\EsriPress\GISTPython\Python Add-Ins\SelectNCount\ReadMe.txt, provided by Esri with the Python Add-in Wizard.

\\EsriPress\GISTPython\Python Add-Ins\addin-assistant.zip, provided by Esri with the Python Add-in Wizard.